Lecture Notes in Artificial Intelligence   9607

Subseries of Lecture Notes in Computer Science

More information about this series at http://www.springer.com/series/1244

Michelangelo Ceci · Corrado Loglisci
Giuseppe Manco · Elio Masciari
Zbigniew W. Ras (Eds.)

New Frontiers in Mining Complex Patterns

4th International Workshop, NFMCP 2015
Held in Conjunction with ECML-PKDD 2015
Porto, Portugal, September 7, 2015
Revised Selected Papers

 Springer

Editors

Michelangelo Ceci
Università degli Studi di Bari Aldo Moro
Bari
Italy

Corrado Loglisci
Università degli Studi di Bari Aldo Moro
Bari
Italy

Giuseppe Manco
ICAR-CNR
Rende
Italy

Elio Masciari
ICAR-CNR
Rende
Italy

Zbigniew W. Ras
University of North Carolina
Charlotte, NC
USA

ISSN 0302-9743   ISSN 1611-3349   (electronic)
Lecture Notes in Artificial Intelligence
ISBN 978-3-319-39314-8   ISBN 978-3-319-39315-5   (eBook)
DOI 10.1007/978-3-319-39315-5

Library of Congress Control Number: 2016939571

LNCS Sublibrary: SL7 – Artificial Intelligence

Printed on acid-free paper

This Springer imprint is published by Springer Nature
The registered company is Springer International Publishing AG Switzerland

New Frontiers in Mining Complex Patterns (NFMCP 2015)

Today, data mining and knowledge discovery are advanced research fields with numerous algorithms and studies to extract complex patterns and models from data in different forms. Although most historical data-mining approaches look for simple patterns in tabular data, there are also numerous recent studies where the focus is on data with a complex structure (e.g., multi-relational data, XML data, Web data, time series and sequences, graphs and trees) with the purpose of extracting complex patterns (e.g., relational patterns, multi-target classification models). Complex data pose new challenges for current research in data mining and knowledge discovery with respect to storing, managing, and mining these sets of complex data.

The 4th International Workshop on New Frontiers in Mining Complex Patterns (NFMCP 2015) was held in Porto in conjunction with the European Conference on Machine Learning and Principles and Practice of Knowledge Discovery in Databases (ECML-PKDD 2015) on September 7, 2015. It aimed at bringing together researchers and practitioners of data mining and knowledge discovery who are interested in the advances and latest developments in the area of extracting nuggets of knowledge from complex data sources.

This book features a collection of revised and significantly extended versions of papers accepted for presentation at the workshop. These papers went through a rigorous review process to ensure compliance with Springer's high-quality publication standards. The individual contributions of this book illustrate advanced data-mining techniques that preserve the informative richness of complex data and allow for efficient and effective identification of complex information units present in such data.

The book is composed of four parts and a total of 15 chapters.

Part I focuses on data stream mining by illustrating some complex predictive or descriptive problems. It consists of six chapters. Chapter 1 is associated with the invited talk given by Jerzy Stefanowski, which discusses the problem of building ensembles in evolving data streams. Chapter 2 compares different algorithms for multi-target regression in data stream mining. Chapter 3 tackles the problem of frequent itemset mining from data streams. Chapter 4 studies methods for discovering organizational structures in event logs, where event logs are considered as streams. Chapter 5, coherently with the invited talk, focuses on the problem of building ensembles in evolving data streams. Chapter 6 presents a method for mining periodic changes in relational data streams.

Part II analyzes issues posed by classification in the presence of complex data either in the input or in the output. It consists of three chapters. Chapter 7 studies the problem of mining classifiers in high-dimensional, imbalanced (and complex) data. Chapter 8 is related to process mining and proposes a new method for classifying (on the basis of the risk) traces of event logs. Chapter 9 presents a new method for learning multi-target predictive clustering trees.

Part III presents algorithms and applications where complex patterns are discovered from complex data. It contains four chapters. Chapter 10 focuses on the problem of using (and generalizing) patterns for cross-domain analogy. Chapter 11 studies the problem of exploiting spectral features extracted from audio for vehicle identification. Chapter 12 presents an algorithm that mines probabilistic frequent subtree kernels. Chapter 13 illustrates a new approach for heterogeneous network decomposition.

Finally, Part IV focuses on sequences. It contains two chapters. Chapter 14 describes a new sequential pattern-mining algorithm that works in the semi-supervised setting. Chapter 15 addresses the problem of forecasting currency exchange rates working on time series.

We would like to thank all the authors who submitted papers for publication in this book and all the workshop participants and speakers. We are also grateful to the members of the Program Committee and external reviewers for their excellent work in reviewing submitted and revised contributions with expertise and patience. We would like to thank Jerzy Stefanowski for his invited talk on "Adaptive Ensembles for Evolving Data Streams — Combining Block-Based and On-Line Solutions." A special thanks is due to both the ECML PKDD workshop chairs and to the ECML PKDD organizers, who made the event possible. We would like to acknowledge the support of the European Commission through the project MAESTRA - Learning from Massive, Incompletely Annotated, and Structured Data (Grant number ICT-2013-612944). Last but not the least, we thank Alfred Hofmann of Springer for his continuous support.

February 2015

Michelangelo Ceci
Corrado Loglisci
Giuseppe Manco
Elio Masciari
Zbigniew W. Ras

Organization

Program Chairs

Michelangelo Ceci	University of Bari Aldo Moro, Italy
Corrado Loglisci	University of Bari Aldo Moro, Italy
Giuseppe Manco	ICAR-CNR, Rende, Italy
Elio Masciari	ICAR-CNR, Rende, Italy
Zbigniew Ras	University of North Carolina, Charlotte, USA & Warsaw University of Technology, Poland

Program Committee

Nicola Barbieri	Yahoo Research London, UK
Elena Bellodi	University of Ferrara, Italy
Petr Berka	University of Economics, Prague, Czech Republic
Ivica Dimitrovski	Ss. Cyril and Methodius University, Skopje, Macedonia
Saso Dzeroski	Jozef Stefan Institute, Slovenia
Bettina Fazzinga	CNR-ICAR Italy Stefano Ferilli, University of Bari, Italy
Dino Ienco	IRSTEA, France
João Gama	University Porto, Portugal
Thomas Gärtner	Fraunhofer Institute IAIS, Germany
Dragi Kocev	Jozef Stefan Institute, Slovenia
Mirco Nanni	KDD-Lab ISTI-CNR Pisa, Italy
Ruggero G. Pensa	University of Turin, Italy
Gianvito Pio	University of Bari, Italy
Chiara Pulice	UNICAL, Italy
Domenico Redavid	Artificial Brain, Italy
Jerzy Stefanowski	Poznan University of Technology, Poland
Herna Viktor	University of Ottawa, Canada
Alicja Wieczorkowska	Polish-Japanese Institute of Information Technology, Poland
Wlodek Zadrozny	University of North Carolina, Charlotte, USA

Additional Reviewers

Sergio Flesca	Felipe Pinagé
Filippo Furfaro	Carlos Ferreira
Douglas Cardoso	Massimo Guarascio
Fabio Fassetti	Ettore Ritacco

Contents

Sequences

Data Stream Mining

Adaptive Ensembles for Evolving Data Streams – Combining Block-Based and Online Solutions

Jerzy Stefanowski[✉]

Institute of Computing Science, Poznań University of Technology,
60-965 Poznań, Poland
Jerzy.Stefanowski@cs.put.poznan.pl

Abstract. Learning ensemble classifiers from concept drifting data streams is discussed. The paper starts with a general overview of these ensembles. Then, differences between block-based and on-line ensembles are examined in detail. We hypothesize that it is still possible to develop new ensembles that combine the most beneficial properties of both types of these classifiers. Two such ensembles are described: Accuracy Updated Ensemble designed to process data blocks and its incremental version, Online Accuracy Updated Ensemble, for learning from single examples.

1 Introduction

The rapid development of the information technology facilities collecting Big Data sets which cause challenges for their storage and processing. Besides massive volumes of data, the characteristics of Big Data includes velocity, variety and complexity of their representation forms. In particular, velocity refers to the speed at which the data is generated and input to analytical systems. It also forces algorithms to process data and produce results in limited time as well as with limited computer resources. These characteristics are especially relevant when data are continuously generated in the form of data streams.

A *data stream* can be viewed as a potentially unbounded sequence of instances which arrive continuously with time-varying intensity. Compared to the static data, mining streams implies new requirements for algorithms, such as constrains on memory usage, restricted processing time, and one scan of incoming examples [19,23,25]. Even more important challenge is that learning algorithms often act in dynamic, non-stationary environments, where the data and target concepts change over time. This phenomena is called *concept drift*. These changes deteriorate the predictive accuracy of the classifiers trained from past data. Examples of real-life concept drifts include spam categorization, weather predictions, monitoring systems, financial fraud detection, and evolving customer preferences, for their review see, e.g. [15,39].

As standard algorithms for supervised classification are not capable to sufficiently tackle evolving data stream requirements, several new algorithms have been recently proposed, for their review see, e.g., [12,14]. Out of several proposals, ensemble methods play an important role in reacting to many types of

© Springer International Publishing Switzerland 2016
M. Ceci et al. (Eds.): NFMCP 2015, LNAI 9607, pp. 3–16, 2016.
DOI: 10.1007/978-3-319-39315-5_1

concept drift [23]. Classifier ensembles are a common way of increasing classification accuracy and due to their modularity they also provide a natural way of adapting to change. In the field of data stream mining, ensemble methods can be divided into block-based and online approaches. Block-based approaches are designed to work in environments were examples arrive in portions, called blocks or chunks, while online ensembles process single instances.

The aims of this study are the following:

1. Briefly survey adaptive ensembles for evolving data streams and stress the main differences between block-based and online solutions.
2. Ask a research question on how the best properties of block and incremental processing can be combined to construct new accurate ensemble classifiers which sufficiently adapt to several types of concept drifts with acceptable memory and time costs.
3. Present two new ensembles being answers to this question: Accuracy Updated Ensemble (AUE) designed to process incoming examples in blocks and its incremental version – Online Accuracy Updated Ensemble (OAUE) processing single examples in the stream.

2   Concept Drift in Data Streams

Learning examples from a stream $\mathcal{S}$ appear incrementally as a sequence of labeled examples $\{\mathbf{x}^t, y^t\}$ for $t = 1, 2, \ldots, T$, where $\mathbf{x}$ is a vector of attribute values and y is a class label ($y \in \{K_1, \ldots, K_l\}$). In this paper, we consider a completely *supervised framework*, where a new example $\mathbf{x}^t$ is classified by a classifier C which predicts its class label. We assume that after some time its true class label y^t is available and the classifier can use it as additional learning information. Other forms of learning are not considered as, e.g., a semi-supervised framework where labels are not available for all incoming examples.

Examples from the data stream can be provided either incrementally (*online*) or in portions (*blocks*). In the first approach, algorithms process single examples appearing one by one in consecutive moments in time, while in the other approach, examples are available only in larger sets called data blocks (or data chunks). Blocks $B_1, B_2, \ldots, B_n$ are usually of equal size and the construction, evaluation, or updating of classifiers is done when all examples from a new block are available. The reader is referred to [3,39] for a more detailed discussion of applications where data are processed more quickly after each instance or in larger portions as blocks. This distinction also refers to availability of class labels. For instance, in some problems data elements are naturally accumulated through some time and labeled in blocks, while access to class labels shortly after making prediction by a classifier (so, online setup) is more demanding.

In case of evolving data streams, target concepts tend to change over time, i.e., the concept, the data is generated from, shifts after a minimum stability period. Concept drift can be defined as follows [14]: In each point in time t, every example is generated by a source with a joint distribution $P^t(\mathbf{x}, y)$ over data; Concepts in data are *stable* if all examples are generated by the same distribution. If for two distinct points in time t and $t + \Delta$ exits such $\mathbf{x}$ that $P^t(\mathbf{x}, y) \neq P^{t+\Delta}(\mathbf{x}, y)$, then concept drift occurs.

Different component probabilities of $P^t(\mathbf{x}, y)$ may change. However in case of supervised classification we are interested in, so called, *real drift*, i.e. posterior probability of classes $P(y|\mathbf{x})$ changes.

Usually two main types of concept drifts are distinguished: *sudden* (abrupt) and *gradual* [34]. The first type of drift occurs when at a moment in time t the source distribution in S^t is suddenly replaced by a different distribution in S^{t+1}. Gradual drifts are not so radical and they are connected with a slower rate of changes which can be noticed while observing a data stream for a longer period of time (e.g., changes of customer preferences). Additionally, some authors distinguish two types of gradual drift [28,37]. The first type of gradual drift refers to the transition phase where the probability of sampling from the first distribution P^j decreases while the probability of getting examples from the next distribution P^{j+1} increases. The other type, called *incremental* (stepwise) drift, may include more sources, however, the difference between them is smaller and the change is noticed only in a longer period of time. In some domains, situations when previous concepts reappear after some time are separately treated and analyzed as *recurrent* drifts [16]. This reoccurrence of drifts could be cyclic (concepts reoccur in a specific order) or not [34]. Moreover, data streams can contain blips (rare events/outliers) and noise, but these are not considered as concept drifts and data stream classifiers should be robust to them.

These types of concept drifts are often considered in the literature. However, some authors study different changes. Typically real concept drift concerns changes for all examples but it could be also a sub-concept change where drift is limited to a subspace of a domain – see discussions on the drift severity in [28,37]. Then, some types of changes are not considered as real drifts. For instance, a *virtual drift* refers to changes in incoming data, e.g. $P(\mathbf{x})$, which do not affect directly $P^j(\mathbf{x}, y)$. Such changes are more difficult to handle, however they are also studied in semi-supervised or unsupervised framework (where labels or incoming examples are not available). One can also observe an increasing interest in studying changes of class probabilities $P(y)$ in data streams or an appearance of novel classes [36]. For more information on these and other changes in underlying data distributions, the reader is referred to wider reviews [14,15,34].

The reaction of classifiers to the above mentioned types of drifts could be easier studied in scenarios with synthetic generated data streams. However, while mining real data streams it is more difficult to detect and identify the type of drift as well as to recognize its characteristic points. We will come back to these issues in the final section.

3   Ensembles for Evolving Data Streams

Although supervised classification has been studied for decades and quite many methods have already been introduced, the concept-drifting data streams require new learning approaches. Looking into the literature reviews, e.g. [12,14], one can notice research on using sliding windows (to manage memory and provide a forgetting mechanism), new kinds of sampling techniques, drift detectors and developing new incremental algorithms. Ensembles are among the most often studied new classifiers. Recall that an *ensemble* (*multiple classifier*) is a set of single classifiers (components) whose decisions are aggregated into a final prediction. The majority voting is a most common aggregation technique [24].

Ensembles are quite naturally adapted to non-stationary data streams. First, due to their modular construction they are flexible to incorporate new data by either introducing a new component into the ensemble or updating knowledge of existing components. Then, forgetting of old knowledge can be done by pruning of too poor performing components. Another option is to continuously change voting weights of component classifiers with respect to recent data elements. Finally, some authors recognize changes in data streams as a mixture of different distributions. It could be modeled as a weighted combination of these distributions – and it is naturally reflected in the structure of voting components inside the multiple classifier.

The comprehensive categorization of ensembles for evolving data streams could be found e.g. in [12,14,23]. For instance, Kuncheva [23] lists the four basic strategies:

- Dynamic combiners – component classifiers are learnt in advance, the ensemble adapts by changing the combination phase (usually by tuning the classifier weights).
- Updating training data – recent data elements are used to online update of component classifiers (e.g. in on-line bagging [30]).
- Updating ensemble members – updating online or retrain in a batch mode.
- Structural changes of the ensemble – replacing "the looser" and adding a new component.

Furthermore, some other authors distinguish between *trigger* vs. *adaptive* (or *active* vs. *passive*) approaches, see e.g. [12].

Active algorithms use special techniques to detect concept drift which trigger changes in classifiers (e.g., rebuilding it from the recent examples). While passive approaches do not contain any drift detector and continuously update the classifier every time a new data element is presented (regardless whether real drift is present in the data stream or not).

Triggers are usually used with single online classifiers. Several drift methods have been recently introduced, see their review in [15]. In case of ensembles they are not applied so frequently. The most well known proposal is an Adaptive Classifier Ensemble (ACE) [29], which includes an online classifier with a drift detector besides several batch learned classifiers. When concept drift is detected, or the number of new incoming examples in the current buffer exceeds the limit, then a new batch component classifier is learnt from the last buffer. The ACE was proposed to overcome the limits of basic block-based ensembles (like SEA [33] or AWE [35]). The experiments carried out in [11,29] have shown that ACE could be more accurate than block based ensembles SEA and AWE. However, our more recent comparative studies [7] have demonstrated that hybrid ensembles, as AUE, or variants of on-line bagging significantly outperformed ACE. Moreover, other experiments with studying different strategies for transforming block-based ensembles into on-line ones [5] have pointed put that incremental updating component classifiers, weighting them with every incoming example and periodical adding a new component to the ensemble are more profitable than incorporating a single drift detector into the structure of the ensemble.

Perhaps triggers should be integrated with ensembles in a more sophisticated way. For instance, in the DDD algorithm [28] a drift detector works with several diversified on-line bagging ensembles. Yet another perspective could be considering multiple detectors, e.g., for each component classifiers (see, e.g., experiences with rule classifiers discussed in [10]).

Adaptive approaches dominate in the current proposals of streaming ensembles. In general, they learn a number of component classifiers on different parts of the data stream Then, they continuously adapt the ensemble and its parameters. Additionally some of them are able to substitute the weakest component with a new candidate classifier learnt from the recent part of the data stream. Such approaches usually cope better with gradual concept drifts as they passively forget old concepts rather than actively detect newest ones.

In this paper we distinguish two general categories of adaptive ensembles:

- *Online ensembles* which incrementally learn/update a classifiers with each single, incoming example in the data stream.
- *Block-based ensembles* which process blocks of data.

Most of block-based ensemble periodically evaluate the component classifiers with the newest data block and substitute the worst ensemble member with a new (candidate) classifier. Practically all proposed approaches work with fixed sized blocks. Below we present the most generic block-based training scheme which will also help us to discuss the new proposed ensembles AUE and OAUE (see the next section).

> **Input**: S: data stream of examples partitioned into blocks of size d,
> k: number of ensemble members, $Q()$: classifier quality measure;
> **Output**: $\mathcal{E}$: ensemble of k weighted classifiers
> **for all** data blocks $B_i \in S$ **do**
> > build and weight candidate classifier C_c using B_i and $Q()$;
> > weight all classifiers C_j in ensemble $\mathcal{E}$ using B_i and $Q()$;
> > **if** $|\mathcal{E}| < k$ **then** $\mathcal{E} \leftarrow \mathcal{E} \cup \{C_c\}$;
> > **else if** $\exists j : Q(C_c) > Q(C_j)$ **then**
> > replace weakest ensemble member with C_c;
> **end for**

For every incoming block B_i, the weights of current component classifiers $C_j \in \mathcal{E}$ are calculated by a quality measure $Q()$. The measure $Q()$ depends on the particular algorithm. In addition to component re-weighting, a candidate classifier C_c is built from the recent block B_i and added to the ensemble if the ensemble's size is not exceeded. If the ensemble is full but the candidate's quality measure is better than at least one of the existing component, the candidate classifier C_c substitutes the weakest ensemble member. Classifying the new coming instance results from the majority of weighted votes of component classifiers. Notice that training component classifiers block-based ensembles often take advantage of batch learning algorithms known from static data mining.

The SEA algorithm [33] was the first of such block-based ensembles and was soon followed by the Accuracy Weighted Ensemble (AWE) [35], which is currently treated as the most representative method of this type. It was also the main inspiration to our research on developing AUE [4]. Another well known proposal is Learn++.NSE [13]. Note that similar block strategies are also used in specialized solutions for reoccurring concepts or inside other proposal for semi-supervised learning from partially labeled streams [27].

In contrast to block-based ensembles, online approaches are designed to act in environments where class labels are available after each incoming example. Usually classifiers are incrementally trained or updated after each instance. Some historical proposals, as WinNow (see its description in [23]), are more focused on changing weights of components and not updating previously generated classifiers. In case of online ensembles there is no common generic training scheme. For instance, DWM and AddExp inherit ideas from the Weighted Majority and pure online learners [21,26]. While Oza's proposal of random updating training sets in online bagging [30] was an inspiration to develop several generalizations, as e.g. Leveraging bagging, ADWIN bagging [1] or a more complex architecture of DDD ensemble [28]. Moreover, the generalized online bagging ensembles are used inside the recent framework for on-line learning from imbalanced data streams [36]. A completely different training scheme is used in other ensembles employing Hoeffding trees, as Option trees (HOT) or Ultra Fast Forest of Trees (UFFT) – see their description in [14]. Moreover, other researches introduced several proposals generalizing online boosting.

For a more precise presentation of various algorithms for learning adaptive ensembles the reader is referred to [3,12,14].

4   AUE and OAUE Ensembles

Block-based ensembles are among the most popular classifiers for concept-drifting data streams. However, their drawback is a delay in reactions to sudden concept drifts, which is caused by analyzing labels only after processing each complete block of examples. Another disadvantage is the difficulty of tuning the block size. Using small-sized blocks could partly help in better reacting to sudden drifts but it will not be reasonable for the periods of stability and it will deteriorate computational costs. One should also notice that in nearly all block-based ensembles component classifiers are trained only once and, then, they are not changed with new blocks until they will be deleted because of too weak predictions.

In case of Accuracy Weighted Ensemble (AWE) some other issues could be questionable, see the discussion in [4]. Here, we will mention only the issues related to our further proposal of ensembles. Recall that the authors of AWE approximate component weights by computing a special version of the mean square error of the component classifier C_j on the recent block B_i. It is defined as

$$MSE_{ij} = \frac{1}{|B_i|} \sum_{\{\mathbf{x},y\} \in B_i} (1 - f_y^j(\mathbf{x}))^2, \tag{1}$$

where $f_y^j(\mathbf{x})$ denotes the probability given by classifier C_j that $\mathbf{x}$ is an instance of class y. Notice that instead of single class predictions, probabilities of all classes are considered. MSE_{ij} is compared to the error of random classifier on the same block (defined as $MSE_r = \sum_y p(y)(1 - p(y))^2$). The voting weight of a component classifier C_j is defined as $w_{ij} = MSE_r - MSE_{ij}$.

AWE ensemble is pruned if errors of component classifiers are worse than the error of a random classifier. However, we noticed in [4] that sometimes this rule may delete all component classifiers in case of some sudden concept drifts and AWE decreases too much its classification accuracy. Although AWE keeps the additional buffer of some pruned classifiers and checks their usefulness in next iterations – the ensemble may be too small to deal with next fast appearing changes. Another problem concerns an evaluation of the new candidate classifier – it is done by means of the 10-fold cross validation on the latest block, which also increases computational time.

On the other hand, most online ensembles can react much faster to sudden drifts and all their components evolve over time. Unlike the block-based approaches (which often employ standard batch learning from the latest block) they need online incremental algorithms. However, they are usually less accurate then block-based classifiers on data streams with gradual or incremental changes. Moreover, they are characterized by higher computational costs as they require frequent updates of components with each new coming example.

Knowing that both categories of ensembles have certain architectural similarities and some of their properties are complementary, Dariusz Brzezinski and Jerzy Stefanowski have decided to study the following hypothesis:

Is it possible to combine best properties of both categories of ensembles to better react to several types of concept drift with satisfactory memory and time?

As a result of their recent research [5,7,8] Dariusz Brzezinski and Jerzy Stefanowski have introduced two new ensembles: Accuracy Updated Ensemble (AUE) for processing streams in blocks and another incremental version, called the Online Accuracy Updated Ensemble (OAUE), dedicated for online classification of instances in the stream one by one.

4.1   Accuracy Updated Ensemble

While constructing Accuracy Updated Ensemble we have kept the block scheme of constructing component classifiers, periodical evaluation of components, learning new candidate classifier with every new block and using it to substitute the weakest component of the current ensemble.

Following the critical analysis of limitations of AWE and similar ensembles we have decided to introduce elements of incremental updating of component classifiers, which should improve the ensemble reaction to various concept drift as well as should reduce an influence of the block size. It means that components of the ensemble should be incrementally learnt with the portion of examples from the recent block. We have resigned from static batch algorithms and chosen other algorithms which could incrementally process the set of examples (from the latest block). In our previous papers we have used Hoeffding Trees available in MOA framework [2]. In [7] we have investigated different strategies of updating component classifiers. The experiments have shown that all component classifiers should be updated after each incoming block of data.

Other main novel contributions include proposing new weighting functions and analyzing the role of new candidate classifier. We have studied different non-linear formulas to better differentiate classifiers and also resigned from pruning with MSE_r. Based on experimental comparative studies we have chosen the following formula:

$$w_{ij} = \frac{1}{MSE_r + MSE_{ij} + \epsilon} \tag{2}$$

A very small positive value ϵ is added to the equation to avoid division by zero problems.

Then, we propose to treat the newest candidate classifier as the possible perfect one (as it has been trained on the latest block of data and should reflect the most recent changes of concepts). Thus, comparing to existing ensemble components it should make the smallest possible error. We do not take into account this error while defining the weight of the candidate classifier – it is defined as $\frac{1}{MSE_r+\epsilon}$. Additionally, this approach is computationally much cheaper than AWE cross-validation estimation of the new candidate weight.

Following experimental studies we have resigned from maintaining an extra buffer of pruned component classifiers and allow to prune Hoeffding Trees (delete the least recently used leaves of trees) to fulfill memory limits [4, 7]. Some other experimental studies of AUC accuracy vs. computational costs [7] have led us to constructing AUE usually with $k = 10$ components and the block size d around 400–500 examples.

We have also carried out an experimental study comparing AUE with 11 additional state-of-the-art data stream methods, including single classifiers, ensembles, and hybrid approaches, in different scenarios [7]. The obtained results confirm that ensemble approaches that use batch classifiers, such as AWE, may suffer accuracy drops after sudden concept drifts, while classifiers with drift detectors are less accurate on gradually drifting streams. What is more important these experiments demonstrated that AUE can offer very high classification accuracy in environments with various types of drift as well as in static environments. AUE provided best average classification accuracy out of all the tested algorithms, while proving less memory consuming than other ensemble approaches, such as Leveraging Bagging or Hoeffding Option Trees. For precise results and their statistical analysis consult [3, 7].

4.2 Online Accuracy Updated Ensemble

Although our experiments have demonstrated that AUE achieves quite high classification accuracy in data streams with various types of changes and it is sufficiently fast and less memory consuming than the most competitive ensembles, it is still designed to process data in blocks. Updating the component classifiers in AUE is done after each block and these classifiers as well as their weights remain unchanged till the end of the next block. In this sense AUE is rather a hybrid classifier, still based on the block-based strategy with limited elements of incremental updating of classifiers.

However, in many environments class labels are available after each incoming example and pure online classifiers should be used. With processing and learning from single examples arriving online, the online ensembles are potentially faster while reacting to changes. On the other hand, online ensembles are characterized by much higher computational costs then block-based classifiers and the adaptation strategies in most of them require special tuning of parameters. Furthermore, most of them do not introduce new components periodically as it is done for block-based ensembles. Recall the experimental results from [20] showing that the adding of the new classifier may be at least so profitable for adapting to gradual drifts as integrating online classifiers with drift detectors.

These motivations have led us to a question whether to design a completely new online ensemble or to construct it with taking into account some good mechanisms from block-based ensembles. We follow the other direction and claim that the weighting mechanisms as well as component evaluations and periodically created candidate classifiers could be still valuable in online ensembles.

In [5] Brzezinski and Stefanowski experimentally studied different strategies of transforming these mechanisms from block-based algorithms into online generalizations. We proposed and evaluated three strategies: re-weighted component classifier with each new example, a windowing technique, the use of an additional incremental learner, and the use of a drift detector. Experimental results demonstrated that not all strategies were equally effective. We have observed that online component re-weighting is the best transformation strategy in terms of average prequential accuracy. Unfortunately, it is also the most time costly strategy, as it requires estimating each component predictive performance after every example. Moreover, we have noticed that elements of incremental learning are crucial in improving classification accuracy. This is also confirmed by experiments with earlier AUC hybrid solutions.

Although the best proposed strategies could be beneficial for constructing online ensembles, they were computationally too costly. In particular, the online re-weighting strategy too drastically increased processing time. Therefore, we looked for other way of tailoring them to a specific version online ensemble, which tries to incorporate incremental learning with online component re-weighting in a time and memory efficient manner.

As a result Brzezinski and Stefanowski have formulated a new incremental ensembles classifier, called Online Accuracy Updated Ensemble (OAUE), which aims at fulfilling these expectations [8]. Similarly to AUE, it maintains a weighted set of component classifiers and predicts the class of incoming examples by aggregating the predictions of ensemble members using a weighted voting rule. However, after processing a new example, each component classifier is weighted according to its accuracy and incrementally trained. We retain a previous approach from AUE, where every d examples a new candidate classifier is created which substitutes the poorest performing ensemble member. In [8], we use Hoeffding Trees as component classifiers, but it is important to note that one could use any online learning algorithm as a base learner.

The main novelty of the OAUE algorithm is the proposal of a new cost-effective component weighting function, which estimates a classifier error on a window of last seen instances in constant time and memory without the need of remembering past examples. For details of its introduction refer to [3, 8].

We also carried out experimental studies analyzing the effect of using different window sizes. The obtained results showed that the prequential accuracy did not significantly change depending on the window size, but larger windows induced higher time and memory costs.

Finally, we experimentally compared OAUE with four representative online ensembles: the Adaptive Classifier Ensemble, Dynamic Weighted Majority, Online Bagging, and Leveraging Bagging. The obtained results demonstrated that OAUE can offer high classification accuracy in online environments regardless of the existence or type of drift. OAUE provided best average classification accuracy out of all the tested algorithms and was among the least time and memory consuming ones.

5   Final Remarks and Open Issues

This paper concerns learning ensemble classifiers for concept-drifting data streams. The differences between approaches for mining static data and data streams were highlighted. We discussed types of concept drifts that occur in non-stationary environments. Then, a general overview of the current data stream ensembles is presented. We focus our attention on differences between block-based and on-line ensembles with respect to: different reaction to various types of drifts, time and memory requirements and strategies to learn component classifiers.

We have hypothesized that it is still possible to develop new types of ensembles that combine the most beneficial properties of these both types of approaches. In the next part of this paper we have presented experiences from using such two algorithms, recently developed in our research team.

The first algorithm, called Accuracy Updated Ensemble, is a more block-based oriented proposal. It includes elements of incremental updating of component ensembles, a new voting function and several solutions to improve its computational costs. Its experimental evaluation shows that it provides the better classification accuracy than other state-of-the-art algorithms, also with acceptable time and memory usage.

Based on some its elements and conclusions from an experimental study [5] we presented an incremental ensemble, called Online Accuracy Updated Ensemble. Its main novelty is a cost-effective component weighting function, which estimates a classifier error on a window of last seen instances in constant time and memory without the need of remembering past examples. Online Accuracy Updated Ensemble aims at retaining the positive elements of Accuracy Updated Ensemble while adding the capability of processing streams online. This has been also confirmed by its experimental evaluation, where OAUE provided best average classification accuracy out of all the tested algorithms and was among the least time and memory consuming ones.

Finally, let us briefly discuss some open research directions for constructing ensembles from complex data streams:

- Considering issues of concept drifts, still more research are needed to better understand their nature and more deeply some types of them. In particular, it concerns gradual drifts. Although in the controlled setup with synthetic data, researchers usually generate progressive changes of concept probabilities, while mining real world data the gradual changes of probability distributions may go along various paths of concept progressions and with different rates (see a discussion in [37]). It may lead to ambiguity as to defining of gradual drifts as well as it makes more difficult to detect various kinds of this drift. In general, most of current drift detectors are designed for sudden drifts only; cf. [15].
- On contrary to evaluating algorithms with synthetic data sets, several open issues still cover questions like: how to identify characteristic points of changes in real data streams, how to measure drift magnitude (e.g., small, medium or high); see, e.g., discussions in [6,22,31,39].
- New procedures for evaluating algorithms in the presence of different drifts should be still developed. Again, they should address real data, however new ideas on constructing special scenarios of artificially constructed changes by modifying real streams should be also studied, see e.g. works on controlled permutations [38] or studying adaptability of the learning classifiers to moments of changes in [31]. Some other researchers postulate multiple-criteria evaluation (e.g., accuracy vs. ability to quick recovery) [6].
- Most of research on concept drifts concern supervised classification and real concept drift $P(y|\mathbf{x})$. Virtual drifts, as covariance drifts of $P(\mathbf{x})$ (also in the context of unsupervised learning) or changing probabilities of classes $P(y)$ (in particular, for imbalanced evolving data streams [9,17,36]) should be more intensively studied and new approaches to detect such changes have to be developed.
- Some researchers argue for including additional background knowledge into drift adaptation techniques, see a discussion with references in [39]. For instance, taking into account seasonal effects while analyzing electricity benchmark data set nicely illustrates usefulness of this postulate.
- Novel class detection or discovery of more structural changes within the concept are still underestimated and call for new proposals.
- Quite often data streams are affected by additional complexity factors. In particular, this problem concerns class imbalance. Class imbalance is an obstacle

even for learning from static data, as classifiers are biased towards the majority classes and tend to misclassify minority class examples. However, it is even more difficult factor for evolving data stream. The research on new specialized classifiers for imbalanced streams has been just started [17]. Up to now, it has resulted in too small number of proposals, also including ensembles. Nearly all their authors consider drifts with the approximately stable, global imbalance ratio between classes. However, as Yao et al. [36] claim changes of the imbalance ratio with time and swapping the roles of minority and majority classes often occur in the context real data. All these issues require new solutions.

- New drift detectors are necessary for changes in imbalanced data streams. Current proposals, as e.g. [36], are too simple modifications of existing detectors.

- Measuring classifier performance for imbalanced data streams is another open issue, as existing online evaluation measures are rather susceptible to class ratio changes over time. Brzezinski and Stefanowski have recently introduced an efficient algorithm to calculate the area under the ROC (Receiver Operating Characteristics) curve incrementally with forgetting on evolving data streams, by using a sorted tree structure with a sliding window [9]. The area under ROC curve (AUC) is a standard measure for evaluating classifiers in static imbalanced data [18]. However, due to its costly computation, AUC is quite difficult to be directly applied to data streams. Experimental results of [9] show that the new proposed measure, called prequential AUC, and the algorithm are sufficiently fast comparing to existing evaluation procedures. Additionally, we have experimentally verified that prequential AUC could be used as a basis for detecting concept drifts and class distribution changes over time.

- Finally, we can repeat after [22] that real world data streams include more complex and heterogeneous representations than simpler attribute-value vectors discussed in this and many other papers. Moreover, more structured outputs (multi-labeled annotations, hierarchies, etc.) may be also present in some applications. Some other problems require analyzing parallel streams at the same moment [32]. All these data complexities also pose new research challenges.

Acknowledgment. The research on this paper was supported by the Polish National Science Center under grant no. DEC-2013/11/B/ST6/00963. The close co-operation with Dariusz Brzezinski on developing the new AUE and OAUE ensembles, and their experimental evaluation, is also acknowledged.

References

1. Bifet, A., Holmes, G., Pfahringer, B.: Leveraging bagging for evolving data streams. In: Balcázar, J.L., Bonchi, F., Gionis, A., Sebag, M. (eds.) ECML PKDD 2010, Part I. LNCS, vol. 6321, pp. 135–150. Springer, Heidelberg (2010)
2. Bifet, A., Holmes, G., Kirkby, R., Pfahringer, B.: MOA: Massive online analysis. J. Mach. Learn. Res. **11**, 1601–1604 (2010)

3. Brzezinski, D.: Block-based and online ensembles for concept-drifting data streams. Ph.D. Thesis, Poznan University of Technology (2015)
4. Brzeziński, D., Stefanowski, J.: Accuracy updated ensemble for data streams with concept drift. In: Corchado, E., Kurzyński, M., Woźniak, M. (eds.) HAIS 2011, Part II. LNCS, vol. 6679, pp. 155–163. Springer, Heidelberg (2011)
5. Brzezinski, D,. Stefanowski, J.: From block-based ensembles to onlinelearners in changing data streams: if- and how-to. In: Proceedings of the 2012 ECML PKDD Workshop on Instant Interactive Data Mining. http://adrem.ua.ac.be/iid2012/
6. Brzezinski, D., Stefanowski, J.: Classifiers for concept-drifting dat streams: Evaluating things that really matter. In: Proceedings of the ECML PKDD 2013 Workshop on Real-World Challenges for Data Stream Mining (2013)
7. Brzezinski, D., Stefanowski, J.: Reacting to different types of concept drift: The accuracy updated ensemble algorithm. IEEE Trans. Neural Netw. Learn. Syst. **25**(1), 81–94 (2014)
8. Brzezinski, D., Stefanowski, J.: Combining block-based and online methods in learning ensembles from concept drifting data streams. Inf. Sci. **265**, 50–67 (2014)
9. Brzezinski, D., Stefanowski, J.: Prequential AUC for classifier evaluation and drift detection in evolving data streams. In: Appice, A., Ceci, M., Loglisci, C., Manco, G., Masciari, E., Ras, Z.W. (eds.) NFMCP 2014. LNCS, vol. 8983, pp. 87–101. Springer, Heidelberg (2015)
10. Deckert, M.: Incremental rule-based learners for handling concept drift: an overview. Found. Comput. Decis. Sci. **38**(1), 35–65 (2013)
11. Deckert, M., Stefanowski, J.: Comparing block ensembles for data streams with concept drift. In: Pechenizkiy, M., Wojciechowski, M. (eds.) ADBIS 2012. AISC, vol. 185, pp. 69–78. Springer, Heidelberg (2012)
12. Ditzler, G., Roveri, M., Alippi, C., Polikar, R.: Learning in nonstationary environments - a survey. IEEE Comput. Intell. Mag. **10**(4), 12–25 (2015)
13. Elwell, R., Polikar, R.: Incremental learning of concept drift in nonstationary environments. IEEE Trans. Neural Netw. **22**(10), 1517–1531 (2011)
14. Gama, J.: Knowledge Discovery from Data Streams. CRC Publishers, Boca Raton (2010)
15. Gama, J., Zliobaite, I., Bifet, A., Pechenizkiy, M., Bouchachia, A.: A survey on concept drift adaptation. ACM Comp. Surv. **46**(4), 44:1–44:37 (2014)
16. Gomes, J., Gaber, M., Sousa, P., Menasalvas, E.: Mining recurring concepts in a dynamic feature space. IEEE Trans. Neural Netw. Learn. Syst. **25**(1), 95–110 (2014)
17. Hoens, T., Chawla, N.: Learning in non-stationary environments with class imbalance. In: Proceedings of the 18th ACM SIGKDD International Conference Knowledge Discovery Data Mining, pp. 168–176 (2012)
18. Japkowicz, N.: Assessment metrics for imbalanced learning. In: He, H., Ma, Y. (eds.) Imbalanced Learning: Foundations, Algorithms, and Applications, pp. 187–206. Wiley-IEEE Press, New Jersey (2013)
19. Japkowicz, N., Stefanowski, J.: A machine learning perspective on big data analysis. In: Japkowicz, N., Stefanowski, J. (eds.) Big Data Analysis: New Algorithms for a New Society. SBD, vol. 16, pp. 1–31. Springer, Switzerland (2016)
20. Kmieciak, M., Stefanowski, J.: Handling sudden concept drift in Enron message data streams. Control Cybern. **40**(3), 667–695 (2011)
21. Kolter, J., Maloof, M.: Dynamic weighted majority: An ensemble method for drifting concepts. J. Mach. Learn. Res. **8**, 2755–2790 (2007)

22. Krempl, G., Zliobaite, I., Brzezinski, D., Hullermeier, E., Last, M., Lemaire, V., Noack, T., Shaker, A., Sievi, S., Spiliopoulou, M., Stefanowski, J.: Open challenges for data stream mining research. SIGKDD Explor. **16**(1), 1–10 (2014)
23. Kuncheva, L.I.: Classifier ensembles for changing environments. In: Roli, F., Kittler, J., Windeatt, T. (eds.) MCS 2004. LNCS, vol. 3077, pp. 1–15. Springer, Heidelberg (2004)
24. Kuncheva, L.: Combining Pattern Classifiers: Methods and Algorithms, 2nd edn. Wiley, Hoboken (2014)
25. Lemaire, V., Salperwyck, C., Bondu, A.: A survey on supervised classification on data streams. In: Zimányi, E., Kutsche, R.-D. (eds.) eBISS 2014. LNBIP, vol. 205, pp. 88–125. Springer, Heidelberg (2015)
26. Littlestone, N., Warmuth, M.: The weighted majority algorithm. Inf. Comput. **108**(2), 212–261 (1994)
27. Masud, M., Gao, J., Khan, L., Han, J., Thuraisingham, B.: A practical approach to classify evolving data streams: training with limited amount of labeled data. In: Proceedings of the 8th IEEE International Conference on Data Mining, pp. 929–934 (2008)
28. Minku, L., White, A., Yao, X.: The impact of diversity on online ensemble learning in the presence of concept drift. IEEE Trans. Knowl. Data Eng. **22**(5), 730–742 (2010)
29. Nishida, K., Yamauchi, K., Omori, T.: ACE: adaptive classifiers-ensemble system for concept-drifting environments. In: Oza, N.C., Polikar, R., Kittler, J., Roli, F. (eds.) MCS 2005. LNCS, vol. 3541, pp. 176–185. Springer, Heidelberg (2005)
30. Oza, N., Russell, S.: Experimental comparisons of online and batch versions of bagging and boosting. In: Proceedings of the 7th ACM SIGKDD International Conference Knowledge Discovery Data Mining, pp. 359–364. ACM Press (2001)
31. Shaker, A., Hullermeier, E.: Recovery analysis for adaptive learning from non-stationary data streams: Experimental design and case study. Neurocomputing **150**, 250–264 (2015)
32. Spiliopoulou, M., Krempl, G.: Mining multiple threads of streaming data. In: Tutorial at the 17th Pacific-Asia Conference on Knowledge Discovery and Data Mining (PAKDD 2013), Gold Coast, Australia, April 2013. https://kmd.cs.ovgu.de/tutorial_pakdd2013.html
33. Street, N., Kim, Y.: A streaming ensemble algorithm (SEA) for large-scale classification. In: Proceedings of the 7th ACM SIGKDD International Conference on Knowledge Discovery and Data Mining, pp. 377–382 (2001)
34. Tsymbal, A.: The problem of concept drift: definitions and related works, Technical report, Dept. Comput. Sci., Trinity College Dublin (2004)
35. Wang, H., Fan, W., Yu, P., Han, J.: Mining concept-drifting data streams using ensemble classifiers. In: Proceedings ACM SIGKDD International Conference Knowledge Discovery Data Mining, pp. 226–235 (2003)
36. Wang, S., Minku, L., Yao, X.: Resampling-based ensemble methods for online class imbalance learning. IEEE Trans. Knowl. Data Eng. **27**(5), 1356–1368 (2015)
37. Webb, G., Hyde, R., Cao, H., Nguyen, H., Petitjean, F.: Characterizing Concept Drift. arXiv preprint (accepted for publication in journal Data Mining and Knowledge Discovery) (2015). arxiv:1511.03816
38. Zliobaite, I.: Controlled permutations for testing adaptive learning models. Knowl. Inf. Syst. **39**(3), 565–578 (2014)
39. Zliobaite, I., Pechenizky, M., Gama, J.: An overview of concept drift applications. In: Japkowicz, N., Stefanowski, J. (eds.) Big Data Analysis: New Algorithms for a New Society. SBD, vol. 16, pp. 91–114. Springer, Switzerland (2016)

Comparison of Tree-Based Methods
for Multi-target Regression on Data Streams

Aljaž Osojnik[1,2(✉)], Panče Panov[1], and Sašo Džeroski[1,2,3]

[1] Jožef Stefan Institute, Jamova Cesta 39, Ljubljana, Slovenia
{aljaz.osojnik,pance.panov,saso.dzeroski}@ijs.si
[2] Jožef Stefan International Postgraduate School, Jamova Cesta 39,
Ljubljana, Slovenia
[3] Centre of Excellence for Integrated Approaches in Chemistry
and Biology of Proteins, Jamova Cesta 39, Ljubljana, Slovenia

Abstract. Single-target regression is a classical data mining task that is popular both in the batch and in the streaming setting. Multi-target regression is an extension of the single-target regression task, in which multiple continuous targets have to be predicted together. Recent studies in the batch setting have shown that global approaches, predicting all of the targets at once, tend to outperform local approaches, predicting each target separately. In this paper, we explore how different local and global tree-based approaches for multi-target regression compare in the streaming setting. Specifically, we apply a local method based on the FIMT-DD algorithm and propose a novel global method, named iSOUP-Tree-MTR. Furthermore, we present an experimental evaluation that is mainly oriented towards exploring the differences between the local and the global approach.

1 Introduction

A common approach to complex data mining tasks is to transform them into simpler tasks, for which have known methods that produce appropriate solutions. This problem transformation approach has been used to address predictive modelling tasks, such as the multi-label classification and multi-target regression tasks. A multi-label classification task can thus be transformed into a collection of binary classification tasks, while a multi-target regression task can be decomposed into several single-target regression problems.

There are, however, methods that forego the reduction to simpler tasks and tackle the complexity head-on. Specifically, in the case of multi-target regression, methods that consider and predict all of the continuous targets at once have received considerable coverage in the literature [13,20]. Almost exclusively, though, these methods have been introduced in the batch setting.

Recently, the streaming setting is becoming more and more prominent, in large part, due to the ever increasing presence of the Big Data paradigm. The streaming setting emphasizes several of the characteristics of Big Data, i.e., the

M. Ceci et al. (Eds.): NFMCP 2015, LNAI 9607, pp. 17–31, 2016.
DOI: 10.1007/978-3-319-39315-5_2

"V"s of Big Data. Specifically, streaming methods need to tackle Velocity – data arriving with high speed; Volume – potentially unbounded number of data instances; and Variability – potential changes in the data itself.

Methods that address the task of multi-target regression in a streaming setting are few and far between, especially those that predict all of the targets at the same time. In this paper, we present a new tree-based method, named iSOUP-Tree-MTR, capable of addressing the task of multi-target regression in this manner. We compare it to the simpler problem transformation approach of using single-target tree-based methods in a streaming setting and show that the iSOUP-Tree-MTR method has superior performance. Finally, we explore the performance of ensembles, e.g., online bagging [15], when using the iSOUP-Tree-MTR method as a base learner.

The structure of the paper is as follows. First, we present the background and related work (Sect. 2). Next, we present several tree-based approaches for multi-target regression on data streams (Sect. 3). Furthermore, we present the research questions we address and the experimental design employed to answer them (Sect. 4). Finally, we conclude with a discussion of the results (Sect. 5), followed by conclusions, and directions for further work (Sect. 6).

2   Background and Related Work

In this section, we define the multi-target regression task and present the constraints imposed by the streaming context. Additionally, we briefly review the state-of-the art in multi-target regression, both in the batch and in the streaming setting.

2.1   Multi-target Regression

In essence, we can look at the task of multi-target regression as an extension of the single-target regression task. In the later, only one continuous variable needs to be predicted. The task of multi-target regression (MTR) deals with predicting multiple numeric variables simultaneously, or, formally, with making a prediction $\hat{y}$ from $\mathbb{R}^n$, where n is the number of targets for a given instance x from an input space X. To categorize the different approaches to MTR we use the nomenclature introduced by Silla and Freitas [12] for the task of hierarchical multi-label classification. The task of simultaneous prediction of all targets at the same time (the *global* approach) has been considered in the batch setting by Struyf and Džeroski [20]. In addition, Appice and Džeroski [1] proposed a method for stepwise induction of multi-target model trees.

2.2   Data Streams

Unlike the batch context, where a fixed and complete dataset is given as input to a learning method, the streaming context presents several constraints that a stream learning method must consider. The most relevant are [2]:

1. the examples arrive sequentially in a specific order;
2. the number of examples can be arbitrarily large;

3. the distribution of examples need not be stationary; and
4. after an example is processed, it is discarded or archived.

The fact that the distribution of examples is not assumed to be stationary means that methods learning in a streaming context should be able to detect and adapt to changes in the distribution (*concept drift*).

2.3   Multi-target Regression on Data Streams

In the streaming context, some recent work has already addressed the task of single- and multi-target regression. Ikonomovska et al. [11] introduced an instance-incremental streaming tree-based single-target regressor (FIMT-DD) that utilizes the Hoeffding bound. This work was later extended for the task of multi-target regression (FIMT-MT) [10]. However, both of these methods had the drawback of ignoring nominal input attributes. Recently, there has been some theoretical debate whether the use of the Hoeffding bound is appropriate [16], however, a recent study by Ikonomovska et al. [9] has shown that in practice the use of the Hoeffding bound produces good results. Other related work includes an instance-based system for classification and regression (IBLStreams), introduced by Shaker et al. [17], which can be, in principle, used for multi-target regression.

3   Tree-Based Approaches for Multi-target Regression on Data Streams

Generally, the quickest way of solving a complex task, such as multi-target regression, is to transform it into a set of simpler tasks that we know how to solve. In the case of multi-target regression, specifically, this is achieved by training a regressor for each of the targets separately, essentially resulting in a collection/ensemble of regressors. The other option for addressing the task of multi-target regression is to produce a regressor which gives predictions for all of the targets simultaneously.

To distinguish between these two approaches we refer to them as local and global, respectively [12]. Specifically, a method that uses one regressor per target is using the *local approach*, while a method that uses one regressor to predict all of the targets simultaneously is using the *global approach*. Recent studies show, that in the batch case, the global approaches outperform the local ones [13]. Global methods tend to (implicitly) exploit the dependencies between the targets.

In this section, we present several tree-based methods for multi-target regression, which utilize the local approach, as well as the global approach. Tree-based methods are often used, as they generally provide good results in terms of predictive performance, while also yielding interpretable models. Finally, we present a baseline method that can be viewed as both local and global and is highly relevant to the methods introduced below.

3.1   A Local Approach to MTR

One of the best known single-target tree-based regressors in the stream setting is the FIMT-DD method [11]. It is based on the Hoeffding bound, which allows the generalization of observations from small samples to the underlying distribution. Similarly to Hoeffding trees for classification [4], FIMT-DD uses the Hoeffding bound to determine the best splits at each node of the regression tree.

We have re-implemented the FIMT-DD method in the Java-based MOA stream-mining framework [3] and extended it to use adaptive models in the leaves, similarly as Duarte et al. [5]. Specifically, each leaf of the tree contains a perceptron. The perceptron is a linear function of the values of the input attributes x that produces the prediction, i.e., $\hat{y} = w \cdot x + b$, where w and b are a learned weight vector and a constant, respectively.

In the original implementation of Ikonomovska et al. [11], the perceptron was always used to make the prediction. However, the adaptive model records the errors of the perceptrons and compares them to the errors of the mean target predictors, which predict the value of the target by computing the average value of the target over the examples observed in a given leaf. In essence, each leaf has two learners, the perceptron and the target mean predictor. The prediction of the learner that (at a given point in time) has a lower error is then used as a final prediction.

To monitor the errors, we use the faded absolute error which is calculated as

$$fMAE_{learner}(m) = \frac{\sum_{j=1}^{m} 0.95^{m-j}|\hat{y}(j) - y(j)|}{\sum_{j=1}^{m} 0.95^{m-j}},$$

where m is the number of observed examples in a leaf, $\hat{y}(j)$ and $y(j)$ are the predicted and real value for the j-th example, respectively, and $learner \in \{perceptron, targetMean\}$. In essence, the faded error is weighted towards more recent examples. Intuitively, the numerator of the fraction is the faded sum of absolute errors, while the denominator is the faded count of examples. For example, the most recent (m-th) example contributes with a weight of 1, the previous example with weight 0.95, and the first example with weight 0.95^{m-1}. This places an emphasis on the more recent examples and generally benefits the perceptron, as we expect its errors to decrease as it learns the weight vector.

We have also implemented a meta-learning method in MOA that creates a homogeneous ensemble of single-target regressors, one regressor for each target, and combines their single-target predictions into a multi-target prediction in real-time to facilitate the use of FIMT-DD as a multi-target regressor. This combination of methods is referred to as the **Local FIMT-DD** method.

3.2   A Global Approach to MTR

As noted earlier, the global approach has been shown to yield better predictive performance in the case of tree-based methods in the batch setting. This has motivated the introduction of global tree-based methods for data streams, i.e., the FIMT-MT method introduced by Ikonomovska et al. [10]. FIMT-MT extends FIMT-DD by replacing the use of the variance reduction heuristic with the intra-cluster variance reduction heuristic, which captures some of the dependencies of the targets. However, one of the major downsides of the FIMT-MT method, is the fact that it completely ignores nominal input attributes.

We have extended the FIMT-MT method by adding the support for nominal input attributes. In addition, we have also proposed the use of this extension to address other structured output prediction tasks, e.g., multi-label classification [14]. The new method is named incremental Structured Output Prediction Tree for Multi-target Regression (iSOUP-Tree-MTR). This method, as the one before, is implemented in the MOA environment.

In each leaf, the iSOUP-Tree method uses an adaptive multi-target model, which consists of a multi-target perceptron and a multi-target target mean predictor. As in the single-target case, the multi-target perceptron produces the prediction vector as $\hat{y} = Wx + b$, where W is the weight matrix and b is the additive vector of constants. On the other hand, the multi-target target mean predictor computes the prediction as the mean value of each of the targets observed at a given leaf. Individually, these learners can be seen as local, however, in a conjunction with a tree building method, they constitute a global method.

For each target y_i, the errors $fMAE^i_{perceptron}$ and $fMAE^i_{targetMean}$ are recorded and the decision which of the predictions to use is made for each variable separately. Formally, for each $i \in \{1, \ldots, n\}$ the prediction $\hat{y}^i_{perceptron}$ is used when $fMAE^i_{perceptron} < fMAE^i_{targetMean}$, otherwise we use $\hat{y}^i_{targetMean}$. Consequently, a final prediction $\hat{y} = (\hat{y}^1, \ldots, \hat{y}^n)$ can contain predictions made by the perceptron (for some targets), the target mean predictor (for other targets), or both.

3.3   Ensemble of Trees for MTR on Data Streams

In this paper, we also consider constructing ensembles of trees for multi-target regression in the stream setting. For this purpose, we use iSOUP-Trees, discussed in the previous subsection, as base learners. To construct the ensemble, we use the bagging method for introducing diversity among the ensemble members. The bagging method for data streams was introduced by Oza et al. [15] and incorporates a probabilistic variation of how many times each given example is "seen" by each of the base learners. In this paper, we refer to this ensemble method as **iSOUP-Tree-MTR bagging**.

3.4   Baseline Method

For the comparison of tree-based methods for multi-target regression on data streams, we use an adaptive multi-target model as a **baseline** regressor. The baseline regressor conveniently corresponds to both an ensemble of leaves using the local approach, as well as to a single leaf in the global iSOUP-Tree approach. In essence, the adaptive model corresponds to a tree-based model that is not allowed to grow, i.e., with leaves that are never split. The baseline method is specifically implemented as a stripped down version of iSOUP-Tree-MTR, building a tree that consists of a single leaf node.

4 Experimental Setup

In this section, we first present the experimental questions that we want to answer in this paper. Next, we discuss the evaluation measures used in the experiments and present the experimental methodology. Finally, we describe the datasets and conclude with the methods used in the experiments.

4.1 Experimental Questions

The first experimental question, that we wish to address in this paper, is the experimental comparison of local and global approaches. As the streaming context imposes several constraints on the learning process, it is not immediately clear whether the findings from the batch setting will be replicated in the streaming setting. While we are specifically using tree-based methods for multi-target regression on data streams, showing that the global approach increases predictive performance could also suggest that this may be generally true, i.e., applicable to other types of methods and structured outputs in the streaming setting.

When the product of the number of input attributes and target variables, which generally corresponds to the time and memory complexity, is low, we expect the local approach to be competitive with the global approach in terms of time and memory. However, as this product increases, we expect that the training of several distinct, even if simpler, models takes more time and especially more memory than the training of a single, more complex, model.

In a single-target study, Ikonomovska et al. [9] have shown no particular differences in the predictive performance of the basic method and the bagging method (therein referred to as FIMT-DD and OBag, respectively). In this work, we wish to investigate whether similar conclusions can be drawn in the multi-target case. To that end, we explore the differences in predictive performance between the iSOUP-Tree-MTR and the iSOUP-Tree-MTR bagging methods. The resource consumption of the bagging method is trivially extrapolated from the resource consumption of a single tree and the number of trees used. Consequently, for this experimental question we focus on the predictive performance.

4.2 Evaluation Measures and Experimental Methodology

To evaluate the predictive performance we define the *average relative mean absolute error* ($\overline{RMAE}$) on an evaluation dataset D as

$$\overline{RMAE} = \frac{1}{n} \sum_{i=1}^{n} \frac{\sum_{j=1}^{|D|} \left| y_j^i - \hat{y}_j^i \right|}{\sum_{j=1}^{|D|} \left| y_j^i - \overline{y}_j^i \right|},$$

where y_j are the true values of the targets, $\hat{y}_j$ are the predictions of the evaluated model and $\overline{y}_j$ are the predictions of a baseline model (all on D). Specifically,

Table 1. Datasets used in the experiments and their properties. N – number of instances, T – number of targets.

Dataset	Domain	N	Input attr.	T
Bicycles [6]	service prediction	17379	12 numeric	3
EUNITE03	quality prediction	8064	29 numeric	5
Forestry Kras [18]	vegetation prediction	60607	160 numeric	11
Forestry Slivnica [19]	vegetation prediction	6218	149 numeric	2
RF1 [21]	environmental prediction	9005	64 numeric	8
RF2 [21]	environmental prediction	7679	575 numeric	8
SCM1d [21]	price prediction	9803	280 numeric	16
SCM20d [21]	price prediction	8966	61 numeric	16

we use the predictions of the baseline perceptron as described in Sect. 3.4 as the baseline (but note that other candidates for the baseline can be used). By definition, this means that the $\overline{RMAE}$ of the baseline is always equal to 1. Essentially, $\overline{RMAE}$ is the relative error averaged over all of the target variables.

To evaluate the time consumption, we will consider the running time of the methods. The memory consumption is measured using the size (in bytes) of the learned models. Both time and memory consumption, as well as $\overline{RMAE}$, are reported at intervals of 1000 examples.

We are using the *prequential* [7] approach for evaluating data stream mining methods. An incoming instance is first used to make a prediction, which is used in the evaluation. Afterwards, the model is updated using the instance. Since the reported errors on data streams can be volatile if reported on an instance by instance basis, due to, e.g., the sampling of different parts of the input space, we calculate the evaluation measures on batches of 1000 examples. Specifically, we calculate the $\overline{RMAE}$ on the first 1000 examples, report it, and then repeat this for examples 1000 through 2000, etc.

4.3 Datasets

We have selected a total of 8 datasets for our experiments based on their size, looking for diversity in the number of input and target attributes. We consider the datasets under the assumption of no concept drift, given that these datasets are generally considered as batch datasets. A summary of the datasets and their properties is shown in Table 1.

The *Bicycles* dataset [6] is concerned with the prediction of demand for rental bicycles on an hour-by-hour basis. The 3 targets represent the number of casual (non-registered) users, the number of registered users and the total number of users for a given hour, respectively.

The *EUNITE03*[1] dataset was used for the competition at the 3rd European Symposium on Intelligent Technologies, Hybrid Systems and their implementation

---

[1] http://www.eunite.org/eunite/news/Summary%20Competition.pdf.

on Smart Adaptive Systems in 2003. The data describes a complex process of continuously manufactured glass products, i.e., the input attributes describe various influences which can or can not be changed by an operator, while the target attributes describe the glass quality.

The data used in the *Forestry Kras* dataset [18] was constructed from multispectral multi-temporal Landsat satellite images and 3D LiDAR recordings of a part of the Kras region in Slovenia. The input attributes and target variables were extracted from the readings, specifically for spatial units of 25 m by 25 m. For specifics on the data preparation procedure, see Stojanova et al. [18]. The task is to predict 11 target variables which correspond to properties of the vegetation in the observed spatial unit.

The data in the *Forestry Slivnica* [19] dataset was, as in the previous case, constructed from multi-spectral multi-temporal Landsat satellite images and 3D LiDAR recordings of a part of the Slivnica region in Slovenia. In this dataset, the task is to predict only 2 target variables: vegetation height and canopy cover.

The river flow datasets, *RF1* and *RF2* [21], concern the prediction of river network flows for 48 h at 8 locations on the Mississippi River network. Each data example comprises the latest observations for each of the 8 locations as well as time-lagged observations from 6, 12, 18, 24, 36, 48 and 60 h in the past. In RF1, each location contributes 8 input attributes, for a total of 64 input attributes and 8 target variables. The RF2 dataset extends RF1 with the precipitation forecast information for each of the 8 locations and 19 other meteorological sites. Specifically, the precipitation forecast for 6 h windows up to 48 h in the future is added, which nets a total of 280 input attributes.

The *SCM1d* and *SCM20d* are datasets derived form the Trading Agent Competition in Supply Chain Management (TAC SCM) conducted in July 2010. The preparation (preprocessing) of the datasets is described by Xioufis et al. [21]. The data instances correspond to daily updates in a tournament – there are 220 days in each game and 18 games per tournament. The 16 targets are the predictions of the next day and the 20 day mean price for each of the 16 products in the simulation, for the SCM1d and SCM20d datasets, respectively.

The Bicycles dataset is available at the UCI Machine Learning Repository[2] and the RF1, RF2, SCM1d and SCM20d datasets are available at the Mulan multi-target regression dataset repository[3]. The examples with missing values in the RF1 and RF2 datasets were removed, so the resulting datasets were somewhat smaller than reported in the repository.

4.4   Compared Methods

For our experiments, we consider the local and global tree-based methods described in Sect. 3. Specifically, we consider the multi-target perceptron as a baseline, the local FIMT-DD-based approach to MTR, the global iSOUP-Tree-MTR approach, and iSOUP-Tree-MTR bagging.

---

[2] https://archive.ics.uci.edu/ml/datasets/Bike+Sharing+Dataset.

[3] http://mulan.sourceforge.net/datasets-mtr.html.

The FIMT-DD method is capable of detecting changes in the concept and adapting to them. However, as this study is oriented towards comparing the local and global tree-based approaches on equal grounds, the change detection and adaptation mechanisms in FIMT-DD have been disabled for this study.

When comparing the bagging approach we also study the effect of increasing the number of trees in the ensemble. To do this we train an ensemble of 100 trees, then use an increasing number of trees for prediction (20, 40, 60, 80 and 100 trees). This allows us to directly see the improvement provided by additional trees, as the results for the higher number of trees are in essence an extension of those with a lower number of trees.

5   Results

In this section we present and discuss the results of our experiments and provide insights into several of the observed phenomena. We consider the performance of the compared methods on the 8 datasets in terms of the evaluation measures: predictive performance and time/memory consumption.

5.1   Predictive Performance ($\overline{RMAE}$)

The predictive performance of the iSOUP-Tree-MTR method appears to be generally worse than that of the Local FIMT-DD method (Fig. 1), excluding the results on the EUNITE03 and SCM20d datasets (Fig. 1b and h). However, we observe that for some datasets (RF1, RF2, SCM1d and Forestry Kras) the iSOUP-Tree-MTR method initially has an identical performance as the baseline. This occurs due to the splitting mechanism of the iSOUP-Tree-MTR method: the root node of the tree is not split until a large number of examples accumulates (in our specific results, 3400 examples on all of the affected datasets).

The affected datasets have a high number of input attributes and/or target variables. In the case of a large number of input attributes, many of them can and do have similar values of the heuristic function so the splitting mechanism cannot easily and quickly determine the best candidate among them. On the other hand, when the number of target variables is high, the aggregation part of the heuristic removes the specificity to particular output attributes, again resulting in similar evaluations of different input attributes. The number of input attributes at which this occurs is much higher (100+) than the number of target variables at which it occurs (already at 8 targets, Fig. 1e and f). As examples accumulate, a tie threshold is reached and a split is made with lower confidence (for details on the tie breaking mechanism, see Ikonomovska et al. [8]). On the above datasets, inappropriate splits are apparently often selected and the performance suffers.

This problem affects the local method as well, on a tree by tree basis for each tree predicting a single-target. However, it is generally easier to distinguish among candidate splits when considering only one target. A potential workaround to this problem is the use of option trees [9], which bypass the myopia of the greedy tree building approach. This problem also naturally affects the results of the ensemble method, in a slightly different way as described below.

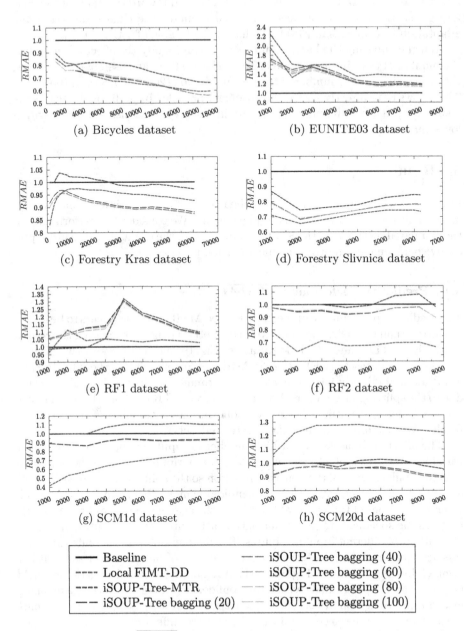

Fig. 1. The evolution of $\overline{RMAE}$ on the selected datasets. Horizontal axes show number of processed examples.

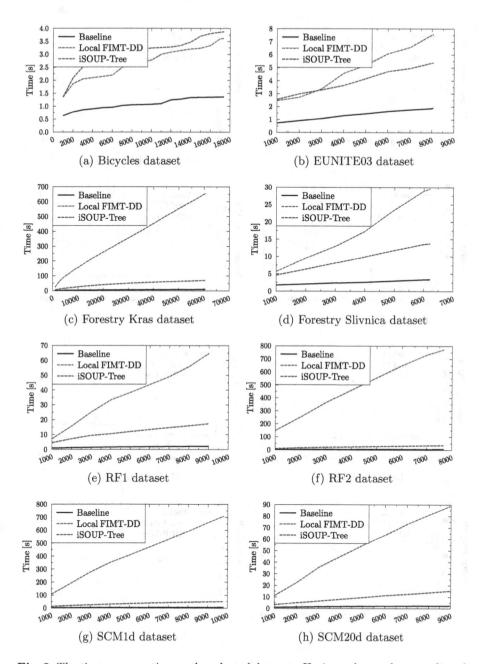

Fig. 2. The time consumption on the selected datasets. Horizontal axes show number of processed examples.

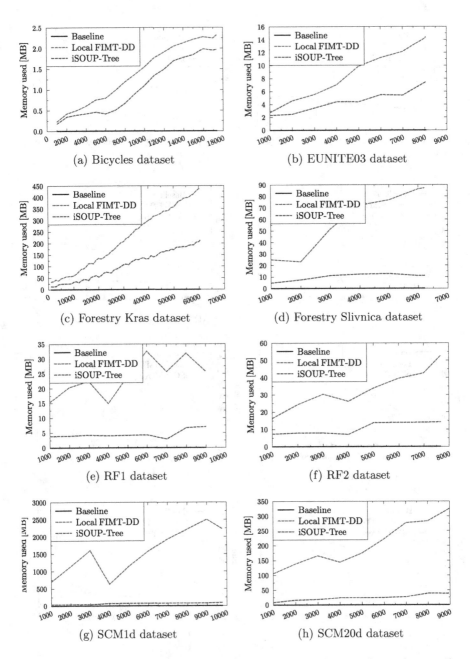

Fig. 3. The memory consumption on the selected datasets. Horizontal axes show number of processed examples.

On the datasets where this does not occur, iSOUP-Tree-MTR and Local FIMT-DD have comparable results. Specifically, Local FIMT-DD outperforms iSOUP-Tree-MTR on the Bicycles and Forestry Kras datasets, while the opposite is true on the EUNITE03 and SCM20d datasets.

When comparing the performance of bagging versus a single tree, bagging generally outperforms the single tree and in some cases even the local method. Interestingly, the variation introduced by the bagging mechanism is sufficient to produce splits (for some of the trees in the ensemble), even for the datasets where iSOUP-Tree-MTR gets "stuck". Therefore, the increase in predictive performance is highest on those datasets. In addition, increasing number of trees generally does not increase the predictive performance much, i.e., the predictive performance of 20 tree ensembles is similar to that of 100 tree ensembles.

5.2   Time Consumption

The results about time consumption (see Fig. 2) are clear. When the number of input attributes and target variables is low, i.e., on the Bicycles and the EUNITE03 dataset, the local FIMT-DD uses less or a similar amount of time as compared to the global iSOUP-Tree-MTR. In these cases, the number of local trees that are traversed and processed is relatively low. In all other cases, iSOUP-Tree-MTR convincingly outpaces the local method, clearly demonstrating that it is a much more scalable approach in terms of time consumption.

5.3   Memory Consumption

The graphs in Fig. 3 indicate that memory consumption of the observed methods is less stable than those of the time consumption. However, the results are more straightforward. The decreases in memory usage occur when a memory intensive leaf node (which stores all the necessary statistics used to evaluate potential splits) is replaced with a split node and new, empty leaf nodes. The iSOUP-Tree-MTR method, which produces only one tree, uses much less memory. As with time, iSOUP-Tree-MTR is more scalable in terms of memory.

6   Conclusions and Further Work

We have conducted an experimental study comparing different approaches for multi-target regression on data streams. The main comparison of this paper considers the local approach (using the FIMT-DD single-target regressor) versus the global approach (using the iSOUP-Tree-MTR multi-target regressor. Additionally, we have compared the performance of iSOUP-Tree-MTR and an ensemble of such trees learned using the online bagging learning method.

Unlike the previous work by Ikonomovska et al. [10], our experiments have highlighted a vulnerability of the greedy, myopic tree building mechanism. Unable to distinguish the best candidate split, the learning method is placed in a staying pattern, where it waits for more information. It finally selects the currently best evaluated split, even though there is not enough support to do so.

More specifically, this problem occurs when the number of input attributes and/or target variables is high. In the first case, when the number of input attributes is high, it is difficult to distinguish them in terms of the employed heuristic. In the second case, when the number of target variables is high, the heuristic values across target variables, results in similar values of the heuristic for many input attributes. The number of input attributes when this occurs is much higher (100+) than the number of target variables (where this phenomena occurs already at 8 targets). This results in bad predictive performance, which occurs on 4 out of 8 of the observed datasets.

On the remaining 4 datasets, we observe that the compared local FIMT-DD and global iSOUP-Tree-MTR have similar results in terms of predictive performance. However, the global method is much more scalable in terms of time and memory consumption. As far as resource consumption is concerned, the local method only outperforms the global one on one dataset – the one with the smallest product of the number of input attributes and target variables.

Furthermore, we have observed that the bagging method generally produces better results in terms of predictive performance. Due to the randomness of the sampling process, it often avoids being trapped into the staying pattern of the single-tree approaches. Note that, the increases in performance due to increasing the number of trees in the ensemble (20 to 100) seem to be negligible.

While the increase in predictive performance brought by the bagging approach can be valuable, it comes at a great cost to both time and memory consumption. While we omitted the specific analysis of resource consumption for the ensemble method, the time and memory demands are linear in the number of trees. This leads to a trade-off between predictive performance and resource usage, if parallelization is not used.

Overall, the global approach does have merit in the streaming context and can produce similar results as the local approach, with a significantly lower resource use footprint. In our future work, we plan to address the problem of insufficient statistical proof when constructing the tree, possibly by implementing option trees [9] within the global iSOUP-Tree-MTR approach. These have been shown to grow faster with fewer available examples, by considering and using multiple attributes for splits, especially in learning the upper levels of the tree.

Acknowledgments. The authors are supported by The Slovenian Research Agency (Grant P2-0103 and a young researcher grant) and the European Commission (Grants ICT-2013-612944 MAESTRA and ICT-2013-604102 HBP).

References

1. Appice, A., Džeroski, S.: Stepwise induction of multi-target model trees. In: 18th European Conference on Machine Learning, pp. 502–509 (2007)
2. Bifet, A., Gavaldà, R.: Adaptive learning from evolving data streams. In: 8th International Symposium on Advances in Intelligent Data Analysis, pp. 249–260 (2009)
3. Bifet, A., Holmes, G., Kirkby, R., Pfahringer, B.: MOA: Massive online analysis. J. Mach. Learn. Res. **11**, 1601–1604 (2010)

4. Domingos, P., Hulten, G.: Mining high-speed data streams. In: 6th ACM SIGKDD, pp. 71–80. ACM, New York (2000)
5. Duarte, J., Gama, J.: Ensembles of adaptive model rules from high-speed data streams. In: 3rd International Workshop on Big Data, Streams and Heterogeneous Source Mining, pp. 198–213 (2014)
6. Fanaee-T, H., Gama, J.: Event labeling combining ensemble detectors and background knowledge. Prog. Artif. Intell. **2**, 1–15 (2013)
7. Gama, J.: Knowledge Discovery from Data Streams. CRC Press, Boca Raton (2010)
8. Ikonomovska, E., Gama, J.: Learning model trees from data streams. In: Boulicaut, J.-F., Berthold, M.R., Horváth, T. (eds.) DS 2008. LNCS (LNAI), vol. 5255, pp. 52–63. Springer, Heidelberg (2008)
9. Ikonomovska, E., Gama, J., Džeroski, S.: Online tree-based ensembles and option trees for regression on evolving data streams. Neurocomputing **150**, 458–470 (2015)
10. Ikonomovska, E., Gama, J., Džeroski, S.: Incremental multi-target model trees for data streams. In: 2011 ACM Symposium on Applied Computing, pp. 988–993. ACM, New York (2011)
11. Ikonomovska, E., Gama, J., Džeroski, S.: Learning model trees from evolving data streams. Data Min. Knowl. Disc. **23**(1), 128–168 (2011)
12. Silla Jr., C.N., Freitas, A.A.: A survey of hierarchical classification across different application domains. Data Min. Knowl. Disc. **22**(1–2), 31–72 (2011)
13. Kocev, D., Vens, C., Struyf, J., Džeroski, S.: Tree ensembles for predicting structured outputs. Pattern Recognit. **46**(3), 817–833 (2013)
14. Osojnik, A., Panov, P., Džeroski, S.: Multi-label classification via multi-target regression on data streams. In: Japkowicz, N., Matwin, S. (eds.) DS 9356. LNCS, vol. 9356, pp. 170–185. Springer, Heidelberg (2015). doi:10.1007/978-3-319-24282-8_15
15. Oza, N.C., Russel, S.J.: Experimental comparisons of online and batch versions of bagging and boosting. In: 7th ACM SIGKDD International Conference on Knowledge Discovery and Data Mining, pp. 359–364. ACM, New York (2001)
16. Rutkowski, L., Pietruczuk, L., Duda, P., Jaworski, M.: Decision trees for mining data streams based on the McDiarmid's bound. IEEE Trans. Knowl. Data Eng. **25**(6), 1272–1279 (2013)
17. Shaker, A., Hüllermeier, E.: IBLStreams: a system for instance-based classification and regression on data streams. Evol. Syst. **3**(4), 235–249 (2012)
18. Stojanova, D., Panov, P., Gjorgjioski, V., Kobler, A., Džeroski, S.: Estimating vegetation height and canopy cover from remotely sensed data with machine learning. Ecol. Inform. **5**(4), 256–266 (2010)
19. Stojanova, D.: Estimating Forest Properties from Remotely Sensed Data by using Machine Learning. Master's thesis, Jožef Stefan International Postgraduate School, Ljubljana, Slovenia (2009)
20. Struyf, J., Dzeroski, S.: Constraint based induction of multi-objective regression trees. In: 4th International Workshop on Knowledge Discovery in Inductive Databases, pp. 222–233 (2005)
21. Xioufis, E.S., Groves, W., Tsoumakas, G., Vlahavas, I.P.: Multi-label classification methods for multi-target regression. CoRR abs/1211.6581 (2012). http://arxiv.org/abs/1211.6581

Frequent Itemsets Mining in Data Streams Using Reconfigurable Hardware

Lázaro Bustio[1,2(✉)], René Cumplido[2], Raudel Hernández[1], José M. Bande[1], and Claudia Feregrino[2]

[1] Advanced Technologies Application Center, 7ª ♯ 21812 e/218 y 222, Rpto. Siboney, Playa, CP 12200 Havana, Cuba
{lbustio,rhernandez,jbande}@cenatav.co.cu
[2] National Institute for Astrophysics, Optics and Electronic, Luis Enrique Erro No 1, Sta. Ma. Tonantzintla, 72840 Puebla, Mexico
{lbustio,rcumplido,cferegrino}@ccc.inaoep.mx

Abstract. Data streams are unbounded and infinite flows of data arriving at high rates which cannot be stored for offline processing. Because of this, classical approaches for Data Mining cannot be used straightforwardly in data stream scenario. This paper introduces a single-pass hardware-based algorithm for frequent itemsets mining on data streams that uses the top-k frequent 1-itemsets. Experimental results of the hardware implementation of the proposed algorithm are also presented and discussed.

Keywords: Data mining · Frequent itemsets mining · Data streams · Reconfigurable hardware · Parallel algorithms

1 Introduction

Data Mining is a research area that investigates the tools and techniques needed to efficiently extract information from large volumes of data. One particularly important technique in Data Mining is frequent itemsets mining aimed at discovering those sets of items that can be found together more than a given number of occurrences in a data set (named *frequent itemsets*) [1].

One scenario that is gaining attention is data streams mining. Frequent itemsets mining on data streams is incipient and the majority of the developed algorithms for this task cannot deal with data streams in an exhaustive fashion because of high incoming rates of data, short processing times needed, and the impossibility of storing the incoming stream. Data streams mining can be found in video and audio streams, network traffic and commercial transactions among others [13]. Such applications need to operate at high speed, so some hardware-based approaches have been proposed to achieve that goal. In such approaches, custom hardware architectures were used as processing platforms due to their capacity to exploit parallelism.

In this paper, a parallel algorithm for frequent itemsets mining on data streams (which was designed to be implemented and executed in hardware) that

© Springer International Publishing Switzerland 2016
M. Ceci et al. (Eds.): NFMCP 2015, LNAI 9607, pp. 32–45, 2016.
DOI: 10.1007/978-3-319-39315-5_3

uses a Landmark Window Model [9] is proposed. This algorithm is based on a systolic tree which is well suited for data streams handling. Also a pre-processing stage that allows to obtain the most important itemsets and to ensure an optimal use of the hardware resources available on the selected hardware device is introduced. In this pre-processing stage, the top-k frequent items are detected and used as starting point for the mining process. To prove the feasibility of the proposed method, several experiments were conducted showing that it outperforms several well known software-based implementations reported in the state-of-the-art which were selected as baseline.

This paper is structured as follows: in Sect. 2 the theoretical basis that support this research is presented. A review of state-of-the-art is addressed in Sect. 3 while Sect. 4 introduces the proposed algorithm. Results are shown in Sect. 5 and finally, conclusions are given in Sect. 6.

2   Theoretical Basis

Let $I = \{i_1, i_2, .., i_n\}$ be a set of n items and T be a transactional dataset[1]:

Definition 1 (Itemset). *An itemset X is a set of items over I such that $X \subseteq I$.*

Definition 2 (Transaction). *A transaction $t \in T$ over I is a couple $t = (tid, X)$ where tid is the transaction identifier, and $X \subseteq I$.*

Definition 3 (Support). *The support of an itemset X is the fraction of transactions in T containing X.*

An itemset is called *frequent* if its support is greater than a given minimal support threshold *minsup*.

Definition 4 (Data stream). *A data stream is a continuous, unbounded and not necessarily ordered, real-time sequence of data items.*

In data stream three main characteristics are presented [2,13,27]:

- *Continuity.* Items in stream arrive continuously at a high rate.
- *Expiration.* Items can be accessed and processed just once.
- *Infinity.* The total number of data is unbounded and potentially infinite.

Definition 5 (Window). *A window in a data stream is an excerpt of transactions.*

Windows can be constructed using one of following approaches [9,17]:

- *Landmark window model.* This model employs some point (called landmark) to start recording where a window begins. The support count of an itemset i is the number of transactions containing i between the landmark and the current time (see Fig. 1a).

---

[1] Dataset is referred to databases, unstructured data file, relational databases or any other data source. In this paper, dataset is used to refer data streams.

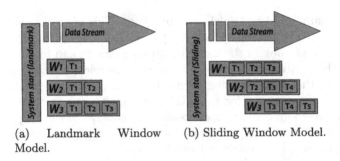

(a)   Landmark   Window
Model.

(b) Sliding Window Model.

Fig. 1. Different windows model. In both figures, three landmarks are presented and in each of them, the processing window is described using the number of transactions that were included.

– *Sliding window model.* This model uses the latest W transactions in the mining process. As newest transactions arrive, the oldest transactions in the sliding windows are excluded. This model can be compared with a FIFO queue. The use of this model imposes a restriction: as some transactions will be included in the mining process, methods for finding expired transactions and discounting the frequency counting of the itemsets involved are required (see Fig. 1b).

In this paper, the Landmark Window Model was selected and used because it can be seen as a general case (and starting point) of others complex windows model such as Sliding, and Landmark is more classical incremental learning instead of a Sliding Window which better deals with evolving characteristics of data streams [9,13,29]. In future works, the Sliding Window Model will be adopted for the mining process.

2.1 Reconfigurable Computing

Reconfigurable Computing is referred to the use of hardware devices in which the functionality of the logic gates is customizable at run-time, and FPGAs are the main exponent of this approach. The architecture of FPGAs is based on a large number of logic blocks which perform basic logic functions. Because of this, FPGAs can implement from a simple logical gate to a complex processing system. FPGAs can be reprogrammed; that is, the circuit can be "erased" and then, a new architecture that implements a new algorithm can be placed. This capability of the FPGAs allows the creation of fully customized architectures, reducing cost and technological risks that are present in traditional circuits designs [10].

3 Related Works

Analyzing the reviewed literature, it can be noticed that there are three preferred approaches: algorithms that use Apriori [1] as the starting point, algorithms that use FP-Growth [14] and those that use Eclat [37]. Algorithms that mimic Apriori [3,4,33,36] implemented in hardware require loading the candidates itemsets and the dataset into the hardware. This strategy is limited by the capacity of the chosen hardware platform. If the number of transactions to manage is larger than the hardware capacity, transactions must be loaded separately in many consecutive times degrading overall performance (the acceleration obtained by the use of FPGA is lost in I/O operations). These issues are forbidden in data streams mining because they do not fulfill the Continuity and Expiration restrictions. One valid option for frequent itemsets mining on data streams is to develop tree-based approaches where transactions flow inside the tree.

Similarly to Apriori-based algorithms, the FP-Growth-based algorithms [22, 24, 30–32] require to copy the mining dataset to the processing platform. They also require to go through the datset in two times, except Mesa et al. [24], however it still needs to download the entire dataset to the hardware device. This is impractical in data streams mining scenario due to the Expiration restriction. Like other reviewed algorithms, authors focused their attention in better data structures that allow the efficient counting of frequent itemsets. However, algorithms based on FP-Growth use the FP-Tree data structure [14] that is based on prefix-trees which are well suited for data streams mining applications.

Eclat-based algorithms [28, 39] use the vertical dataset representation in order to save memory and processing time. In [28, 39], authors use the intersection of items to compute the support. They show that it is more efficient than hash-trees. All the Eclat-based implementations propose an hybrid approach, where the most time and/or memory consuming functions were downloaded to custom hardware while the software controls the execution flow and data structures. Although the vertical dataset representation allows to save memory and processing time, it is not compatible with the Expiration restriction of data streams.

4   A Method for Frequent Itemsets Mining Using Reconfigurable Hardware

The basic idea of the proposed method is to develop a tree structure of processing units where transactions from data streams flow from the root node to the leafs nodes. The tree structure can be seen as a *systolic tree* where each of its nodes has one bottom node (child) and one right node.

Figure 2 shows the systolic tree where vertically-arranged nodes represent a prefix path and parent nodes contain the prefix itemset (that represent transactions of data streams) for their children. Taking a random node r, the sub-tree who had the node r as root is formed by all possible combinations of items with the itemset stored in r as their prefix, this leads to recursive mining strategies. The systolic tree data structure implements a distributed control scheme where processing and control logic are distributed among the nodes. A schematic of the systolic tree is shown in Fig. 3.

Data streams are data sources that usually evolve in time. Because of this, it is unrealistic and meaningless in several applications to determine all frequent itemsets that have been transmitted. Also in such cases, changes of patterns and their trends are more relevant than the patterns themselves. Detecting changes

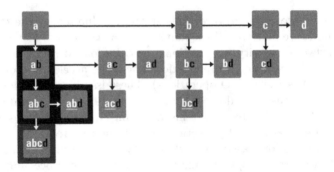

Fig. 2. Systolic tree data structure used in data stream mining. Remarked section shows a sub-tree with nodes that have the itemset *ab* of a transaction *t* as root.

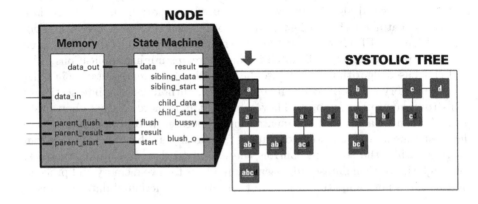

Fig. 3. Schematic of the processing node (gray and magnified at left side of the figure) and structure of the systolic tree (at right side of the figure).

on transmitted patterns in data streams is studied in several ways, and one of these ways is to find the most occurring 1-itemsets (also named *heavy hitters*) [11,20,23,25,26,38]. This problem have been also attacked from the hardware perspective [18,34].

4.1 Frequent 1-Itemsets Detection

In data streams mining the detection of top-frequent 1-itemsets can be seen as a pre-processing stage where the most representative itemsets transmitted in the stream are discovered. Using the top-frequent 1-itemsets as starting point to reduce the search space also allows to reduce the computational resources needed and this ensures to obtain the most valuable frequent itemsets. Also, frequent 1-itemsets detection is useful to verify a concept drift in the knowledge [15,16,35] which is transmitted in the stream. The use of top-frequent 1-itemsets in data stream mining tasks was used in [5,19] and the validity of using top-frequent 1-itemset detection is supported on the "Downward Closure" property [1].

In the reviewed literature, it can be noticed that reported algorithms use the k-first different items received from the data stream (where k is the maximum number of single items that can be assimilated by hardware platforms). These k-first items are taken in arrival order and may not be the most representative of the stream. It may occur that the processing systolic tree deals with items that are not interesting and do not generate any frequent itemsets but nevertheless consume hardware resources. In order to alleviate this situation, in this paper a strategy to ensure the received itemsets will be the most representatives is adopted: this is achieved by obtaining top-k frequent 1-itemsets in each window. This process is performed in the window creation phase and a trade-off must be adopted: to use some transactions (e.g. 1/10 of the window length) to obtain top-k frequent 1-itemsets in order to obtain the most valuable frequent itemsets.

In the proposed processing scheme the top-k frequent 1-itemsets are used, where k is directly related to available hardware resources. Those top-k frequent 1-itemsets are obtained in a pre-processing stage implemented in software. Later, those top-k frequent 1-itemsets are transmitted to the mining process (which is implemented in hardware).

4.2   Proposed Method

Figure 4 shows the proposed processing scheme. It shows the pre-processing stage (software) and the mining process (hardware). Algorithm 1 is executed simultaneously in each node of the systolic tree. For each transaction t in Landmark window when a new itemset X arrives to a node occupied by an item $i_j \in I$, one of following decisions must be taken:

1. If $\{i_j\} \subseteq X$ then the frequency counter of the node is incremented and the itemset $X - \{i_j\}$ is flowed to child and right node.
2. If $\{i_j\} \nsubseteq X$ then X is flowed to right node.

Algorithm 1 simultaneously uses Depth First Search (DFS) and Breadth First Search (BFS) traversals (from lines 20 to 25) allowing a high level of parallelism: two dimensional searches are performed concurrently. These DFS and

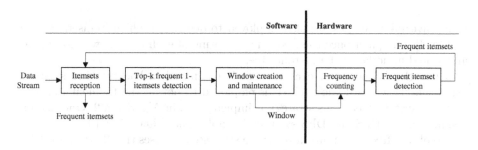

Fig. 4. Processing scheme proposed which is composed by a software-side which executes a frequent 1-itemsets detection and a hardware-side which executes the mining process in the systolic tree.

Algorithm 1. Parallel algorithm for finding the frequency counting of flushed transactions in a Landmark Window.

Input: Landmark Window *landmark_window*
Output: Systolic tree with the counting frequency of each itemset.

1   $n_i \leftarrow$ *systolyc_tree.RootNode*;
2   **foreach** *transaction t in landmark_window* **do**
3     $\lfloor$ Traverse(t, n_i);

4   **Procedure** Traverse(t, n_i);
5   **if** $n_i.IsOccupied == false$ **then**
6      $n_i.IsOccupied = true$;
7      $n_i.Item = t.ItemAt(0)$;
8   FlushInParallel(t, n_i);
9   **else**
10   **if** $t.Contain(n_i.Item) == true$ **then**
11       $\lfloor$ FlushInParallel(t, n_i);
12   **else**
13       $\tilde{n}_i \leftarrow n_i.RightNode$;
14   Traverse$(t, \tilde{n}_i)$;

15   **EndProcedure**;

16   **Procedure** FlushInParallel(t, n_i);
17   $n_i.Counter + +$;
18   $\tilde{t} = t.Exclude(n_i.Item)$;
19   **if** $\tilde{t}.IsEmpty == false$ **then**
20   **StartParallalelBlock::**
21     $\tilde{n}_i \longleftarrow n_i.ChildNode$;
22   Traverse$(\tilde{t}, \tilde{n}_i)$;
23     $\tilde{n}_i \longleftarrow n_i.RightNode$;
24   Traverse$(\tilde{t}, \tilde{n}_i)$;
25   **EndParallelBlock;**;

26   **EndProcedure**;

27   **return** *systolic_tree*;

BFS traversal strategies are also employed to determine which itemsets can be regarded as frequent once the frequency counting of each itemset was computed and stored in nodes of the systolic tree.

After the frequency of each itemset is computed, a backtracking strategy using the "Downward Closure" [1] is employed to obtain the frequent itemsets from the systolic tree. This strategy is implemented in Algorithm 2. Once again, a simultaneous BFS and DFS strategy is implemented lines 7 to 10.

To obtain frequent items, a recursive strategy that uses the "Downward Closure" property [1] stated before is used. In this strategy, if a node is declared as *frequent*, then its child and right nodes must be processed recursively to determine whether they are frequent or not. On the contrary, if a node is regarded

---

Algorithm 2. Parallel algorithm for finding frequent itemsets.

Input: Flushing flag $flush$, Minimum support value min_sup.
Output: Frequent itemsets and their frequency counting
$< itemset, frequency >$.

1 $n_i \leftarrow systolyc\_tree.RootNode$;
2 **if** $(flush == true)$ **then**
3 $\quad$ FlushMethod(n_i, min_sup);

4 **Procedure** FlushMethod(n_i, min_sup)
5 **if** $(n_i.counter \geq min\_sup)$ **then**
6 $\quad$ **if** $(n_i.ChildNode! = null) and (n_i.RightNode! = null)$ **then**
7 $\quad\quad$ **StartParallalelBlock:**
8 $\quad\quad$ FlushMethod($n_i.ChildNode, min\_sup$);
9 $\quad\quad$ FlushMethod($n_i.RightNode, min\_sup$);
10 $\quad\quad$ **EndParallelBlock;**
11 $\quad\quad$ $n_i.GatewayMode = true$;
12 $\quad$ **else**
13 $\quad\quad$ FlushResult.Add($n_i, n_i.Counter$);

14 **else**
15 $\quad$ FlushResult.Add($n_i, n_i.Counter$);

16 **return** $FlushResult$;

---

as *infrequent*, its descendants will be infrequent and the process can be stopped (consequence of the "Downward Closure" property). In this way, the traverse strategy to obtain the frequent itemsets in the systolic tree is optimized.

5 Results

The proposed method was modeled in VHDL using Xilinx ISE Suite 14.2 and targeted for a Virtex 5 XC5VLX330T FPGA device. After the architecture was synthesized and implemented, the occupied area is 90.3 %, holding 1161216 processing nodes. This allows to handle a maximum of 20 different items, while the largest itemsets handled by FPGA-accelerated architecture reported is about 11 items. The maximum operation frequency obtained for this architecture is 238 Mhz. This means that the proposed architecture can process 5.07×10^6 transactions (containing at most 20 items) per second (worst case), this derives in a throughput of 1.19 Gbps. Besides, as far as we know, this is the first FPGA-accelerated architecture for frequent itemset mining over data streams reported.

To validate the performance of the proposed systolic tree architecture, some experiments were conducted simulating data streams with 4 different datasets. MSNBC and Chess datasets were taken from UCI repository [21] and the other two were created with the Almaden IBM Synthetic Dataset Generator [12]. Employed datasets are described in Table 1.

Table 1. Dataset specifications.

Dataset	Size (Mb)	#Trans	#Items	Ave.IT
MSNBC	4.219	989 818	17	2.82
Chess	0.340	3 196	75	37
T10I20D100K	1.468	100 000	20	10
T20I20D1M	27.123	1 000 000	20	20

Since there are no hardware architectures reported for frequent itemsets mining on data streams, the best software implementations (concerning of performance and resources consumption) of Apriori [7], FP-Growth [8] and Eclat [7] algorithms that were reported in the state-of-the-art, were used as baseline. These implementations were downloaded from Borgelt's website [6] and several experiments were conducted using MSNBC, Chess, T10I20D100K and T20I20D1M datasets on an Intel Core i3 at 1.8 GHz and 4 Gb of RAM. Experiments using all datasets were conducted in software using several *minsup* values and timing results were compared versus the same experiments conducted in hardware. Timing results are shown in Fig. 5. These results are shown to give an idea of the performance of the proposed method compared against some well established state-of-the-art algorithms. Window size was set to 100 000 transactions. This allows to handle in one window the MSNBC, Chess and T10I10D100K datasets, while T20I20D1M was separated in 10 windows.

5.1   Discussion

As it is shown in Fig. 5, the proposed architecture outperforms all the baseline algorithms in all datasets. Note that, for the sake of clarity, processing times that exceeds 3 min are not shown. All datasets, except Chess, contain 20 or less single items, so they can be mapped directly into the hardware architecture. Chess dataset contains 75 items so the systolic tree only takes into account for the mining process the most frequent 20 items arrived. In this case, the mining process is approximate, discovering a subset of total frequent itemsets but this subset express more accurately the behavior of frequent itemsets if it were obtained using the first 20 items arrived. Itemsets regarded as frequent in this case have the same frequency counting as if they had been calculated with any baseline algorithm. For MSNBC, T10I20D100K and T20I20D1M, the mining process is exact and it was performed one order of magnitude times faster than baseline algorithms.

Experiments demonstrate that the proposed architecture is insensitive to variations in support threshold: this is explained because of all hardware structures needed for mining incoming data stream is available and it is the same whether the minimum support threshold is 1 % or 99 %. Restrictive support threshold conduce to less frequent itemsets while relaxed support threshold conduce to more frequent itemsets. Performance of software implementations of

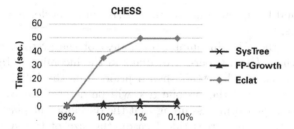

(a) Execution time for the proposed architecture with Chess dataset.

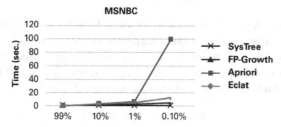

(b) Execution time for the proposed architecture with MSNBC dataset.

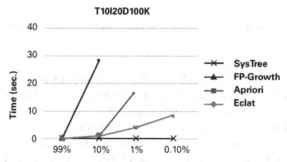

(c) Execution time for the proposed architecture with T20I20D100K synthetic dataset.

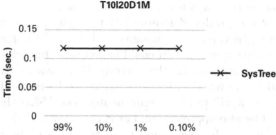

(d) Execution time for the proposed architecture with T10I20D1M synthetic dataset. Here, timing results that exceeds 180 s were dropped from chart.

Fig. 5. Performance evaluation of the proposed algorithm and architecture for different datasets. The X axis represents the variation of the support threshold expressed in %.

Apriori, Eclat and FP-Growth are quite sensitive in its performance to changes on support value.

Theoretically, the size (in number of nodes) of the systolic tree is determined by $|I|$, where I is a super-set of items in the incoming data stream (see Sect. 2). Since $|I|$ cannot be established a priori (due the Infinity property of data streams), the size of the systolic tree will be determined by the capacity of the development platform. Assuming that the development platform contains enough computational resources (ideal case), the size of the systolic tree (in number of nodes) that can hold any streams formed by items of I will be:

$$nodes = 2^{|I|} - 1 \tag{1}$$

In a real case, the available resources of a FPGAs are limited. Supposing 100 % of area occupation in selected hardware device, a number p will be the maximum number of processing nodes that can be supported by the selected device. Then the maximum number max_{items} of items in I in the incoming data stream that can be handled by the chosen device will be:

$$max_{items} = \log_2(p+1) \tag{2}$$

For example, if the selected hardware device can hold 1024 nodes ($p = 1024$) in systolic tree, then the maximum number of items of I that can be handled will be 10 ($|I|$).

It is important to notice that for a certain number n of items in set I, if

$$2^n - 1 > p \tag{3}$$

the systolic tree cannot hold all possible combinations of itemsets and therefore some itemsets will not be taken into account during the mining process. In other words, the number of processing nodes needed to handle all possible combinations of itemsets generated from I exceeds the maximum number of processing nodes that can be mapped into the selected FPGA. Here, mining will be approximate with no false positives produced. In this case, the selection of top-k frequent 1-itemsets allows to obtain the most valuable items to produce frequent itemsets. Obtaining some information about the incoming data stream introduces some delay (which is negligible) in the mining process but it implies a higher quality of the produced itemsets. If the systolic tree is occupied by items in arrival order, it cannot be guaranteed that all of the received items in transaction will produce itemsets that could be frequent. In such case, the use of the systolic tree is not optimal. By the contrary, if the systolic tree is occupied by items that had been proved to be frequent, it can be ensured that all combination of those items will produce frequent itemsets. Using this strategy the mining process will be also approximate, but it is guaranteed that the produced itemsets will be the most frequent ones. From the point of view of users, it is more useful to obtain "some" frequent itemsets than to obtain "all" frequent itemsets [11,23,25]. In this paper, using the top-frequent 1-itemsets it is ensuring that frequent itemsets obtained describe better the data stream behavior than just using the k-first arrived 1-itemsets.

On the contrary, if:

$$2^n - 1 \leq k \tag{4}$$

the systolic tree can hold all the possible combinations of itemsets, therefore the mining process will be exact.

6   Conclusions

This paper introduces a new parallel algorithm for frequent itemsets mining in data streams which is designed to be implemented in a custom hardware architecture. The proposed algorithm is based on a tree data structure that allows to increase the mining performance. The corresponding hardware architecture is based on a systolic tree approach where the control logic is distributed among all processing nodes. Here, the use of top-k frequent 1-itemsets allows to optimize the use of the systolic tree and to obtain the most valuable frequent itemsets. Experimental results showed that the hardware architecture is able to extract correctly all itemsets from data streams with a significant speed up when compared against software-based implementations.

References

1. Agrawal, R., Srikant, R.: Fast algorithms for mining association rules in large databases. In: Proceedings of the 20th International Conference on Very Large Data Bases, VLDB 1994, pp. 487–499. Morgan Kaufmann Publishers Inc., San Francisco (1994)
2. Babcock, B., Babu, S., Datar, M., Motwani, R., Widom, J.: Models and issues in data stream systems. In: Proceedings of the Twenty-first ACM SIGMOD-SIGACT-SIGART Symposium on Principles of Database Systems, PODS 2002, pp. 1–16. ACM, New York (2002)
3. Baker, Z.K., Prasanna, V.K.: Efficient hardware data mining with the Apriori algorithm on FPGAs. In: Proceedings of the 13th Annual IEEE Symposium on Field-Programmable Custom Computing Machines, FCCM 2005, pp. 3–12. IEEE Computer Society, Washington (2005)
4. Baker, Z.K., Prasanna, V.K.: An architecture for efficient hardware data mining using reconfigurable computing systems. In: 14th Annual IEEE Symposium on Field-Programmable Custom Computing Machines, 2006, FCCM 2006, pp. 67–75, April 2006
5. Baralis, E., Cerquitelli, T., Chiusano, S., Grand, A., Grimaudo, L.: An efficient itemset mining approach for data streams. In: König, A., Dengel, A., Hinkelmann, K., Kise, K., Howlett, R.J., Jain, L.C. (eds.) KES 2011, Part II. LNCS, vol. 6882, pp. 515–523. Springer, Heidelberg (2011)
6. Borgelt, C.: Software for frequent pattern mining. http://www.borgelt.net/fpm.html. Accessed on 20 May 2015
7. Borgelt, C.: Efficient implementations of Apriori and Eclat. In: FIMI. CEUR Workshop Proc., vol. 90. CEUR-WS.org (2003)
8. Borgelt, C.: An implementation of the FP-growth algorithm. In: Proceedings of the 1st International Workshop on Open Source Data Mining: Frequent Pattern Mining Implementations, OSDM 2005, pp. 1–5. ACM (2005)

9. Cheng, J., Ke, Y., Ng, W.: A survey on algorithms for mining frequent itemsets over data streams. Knowl. Inf. Syst. **16**(1), 1–27 (2008)

10. Compton, K., Hauck, S.: Reconfigurable computing: A survey of systems and software. ACM Comput. Surv. **34**(2), 171–210 (2002)

11. Cormode, G., Korn, F., Muthukrishnan, S., Srivastava, D.: Finding hierarchical heavy hitters in data streams. In: Proceedings of the 29th International Conference on Very Large Data Bases, VLDB 2003, vol. 29, pp. 464–475. VLDB Endowment (2003)

12. Corporation, I.B.M.: IBM quest market-basket synthetic data generator. http://www.cs.loyola.edu/cgiannel/assoc_gen.html. Accessed on 20 April 2015

13. Golab, L., Ozsu, M.: Data stream management issues - a survey. SIGMOD Rec. **32**(2), 5–14 (2003)

14. Han, J., Pei, J., Yin, Y.: Mining frequent patterns without candidate generation. In: Proceedings of the 2000 ACM SIGMOD International Conference on Management of Data, SIGMOD 2000, pp. 1–12. ACM, New York (2000)

15. Hulten, G., Spencer, L., Domingos, P.: Mining time-changing data streams. In: Proceedings of the Seventh ACM SIGKDD International Conference on Knowledge Discovery and Data Mining, KDD 2001, pp. 97–106. ACM, New York (2001)

16. Jiang, N., Gruenwald, L.: Research issues in data stream association rule mining. SIGMOD Rec. **35**(1), 14–19 (2006)

17. Jin, R., Agrawal, G.: Frequent pattern mining in data streams. In: Aggarwal, C. (ed.) Data Streams. Advances in Database Systems, vol. 31, pp. 61–84. Springer, US (2007)

18. Lai, Y.K., Wang, N.C., Chou, T.Y., Lee, C.C., Wellem, T., Nugroho, H.T.: Implementing on-line sketch-based change detection on a NetFPGA platform. In: 1st Asia NetFPGA Developers Workshop (2010)

19. Lee, W., Stolfo, S.J., Mok, K.W.: Adaptive intrusion detection: A data mining approach. Artif. Intell. Rev. **14**(6), 533–567 (2000)

20. Thanh, L.H., Calders, T.: Mining top-k frequent items in a data stream with flexible sliding windows. In: Proceedings of the 16th ACM SIGKDD International Conference on Knowledge Discovery and Data Mining, KDD 2010, pp. 283–292. ACM, New York (2010)

21. Lichman, M.: UCI machine learning repository (2013). Accessed on 20 April 2015

22. Malarvizhi, S.P., Sathiyabhama, B.: Enhanced reconfigurable weighted association rule mining for frequent patterns of web logs. Int. J. Comput. **13**(2), 97–105 (2014)

23. Manku, G.S., Motwani, R.: Approximate frequency counts over data streams. In: Proceedings of the 28th International Conference on Very Large Data Bases, VLDB 2002, pp. 346–357. VLDB Endowment (2002)

24. Mesa, A., Feregrino-Uribe, C., Cumplido, R., Hernández-Palancar, J.: A highly parallel algorithm for frequent itemset mining. In: Martínez-Trinidad, J.F., Carrasco-Ochoa, J.A., Kittler, J. (eds.) MCPR 2010. LNCS, vol. 6256, pp. 291–300. Springer, Heidelberg (2010)

25. Metwally, A., Agrawal, D.P., El Abbadi, A.: Efficient computation of frequent and top-k elements in data streams. In: Eiter, T., Libkin, L. (eds.) ICDT 2005. LNCS, vol. 3363, pp. 398–412. Springer, Heidelberg (2005)

26. Metwally, A., Agrawal, D., Abbadi, A.E.: An integrated efficient solution for computing frequent and top-k elements in data streams. ACM Trans. Database Syst. **31**(3), 1095–1133 (2006)

27. Pramod, S., Vyas, O.P.: Data stream mining: A review on windowing approach. Global J. Comput. Sci. Technol. **12**(11-C) (2012)

28. Shi, S., Qi, Y., Wang, Q.: Accelerating intersection computation in frequent item-set mining with FPGA. In: 2013 IEEE 10th International Conference on High Performance Computing and Communications 2013, pp. 659–665, November 2013
29. Shie, B.E., Yu, P.S., Tseng, V.S.: Efficient algorithms for mining maximal high utility itemsets from data streams with different models. Expert Syst. Appl. **39**(17), 12947–12960 (2012)
30. Sun, S., Steffen, M., Zambreno, J.: A reconfigurable platform for frequent pattern mining. In: International Conference on Reconfigurable Computing and FPGAs, 2008, ReConFig 2008, pp. 55–60, December 2008
31. Sun, S., Zambreno, J.: Mining association rules with systolic trees. In: International Conference on Field Programmable Logic and Applications, 2008, FPL 2008, pp. 143–148, September 2008
32. Sun, S., Zambreno, J.: Design and analysis of a reconfigurable platform for frequent pattern mining. IEEE Trans. Parallel Distrib. Syst. **22**(9), 1497–1505 (2011)
33. Thoni, D.W., Strey, A.: Novel strategies for hardware acceleration of frequent item-set mining with the Apriori algorithm. In: International Conference on Field Pro-grammable Logic and Applications, 2009, FPL 2009, pp. 489–492, August 2009
34. Tong, D., Prasanna, V.: Online heavy hitter detector on FPGA. In: 2013 Interna-tional Conference on Reconfigurable Computing and FPGAs (ReConFig), pp. 1–6. IEEE (2013)
35. Wang, H., Fan, W., Yu, P.S., Han, J.: Mining concept-drifting data streams using ensemble classifiers. In: Proceedings of the Ninth ACM SIGKDD International Conference on Knowledge Discovery and Data Mining, KDD 2003, pp. 226–235. ACM, New York (2003)
36. Wen, Y.H., Huang, J.W., Chen, M.S.: Hardware-enhanced association rule mining with hashing and pipelining. IEEE Trans. Knowl. Data Eng. **20**(6), 784–795 (2008)
37. Zaki, M.J.: Scalable algorithms for association mining. IEEE Trans. Knowl. Data Eng. **12**(3), 372–390 (2000)
38. Zhang, Y., Singh, S., Sen, S., Duffield, N., Lund, C.: Online identification of hier-archical heavy hitters: Algorithms, evaluation, and applications. In: Proceedings of the 4th ACM SIGCOMM Conference on Internet Measurement, IMC 2004, pp. 101–114. ACM, New York (2004)
39. Zhang, Y., Zhang, F., Jin, Z., Bakos, J.D.: An FPGA-based accelerator for frequent itemset mining. ACM Trans. Reconfigurable Technol. Syst. **6**(1), 2:1–2:17 (2013)

Discovering and Tracking Organizational Structures in Event Logs

Annalisa Appice[✉], Marco Di Pietro, Claudio Greco, and Donato Malerba

Dipartimento di Informatica, Università degli Studi di Bari Aldo Moro,
via Orabona, 4, 70126 Bari, Italy
{annalisa.appice,donato.malerba}@uniba.it

Abstract. The goal of process mining is to extract process-related information by observing events recorded in event logs. An event is an activity initiated or completed by a resource at a certain time point. Organizational mining is a subfield of process mining that focuses on the organizational perspective of a business process. It considers the resource attribute and derives a profile that characterizes the behavior of a resource in a specific business process. By relating resources associated with correlated profiles, it is possible to define a social network. This paper focuses on the idea of performing organizational mining of event logs via social network mining. It presents a framework that resorts to a stream representation of an event log. It adapts the time-based window model to process this stream, so that window-based social resource networks can be constructed, in order to represent interactions between resources operating at the data window level. Finally, it integrates specific algorithms, in order to discover (overlapping) communities of resources and track the evolution of these communities over consecutive windows. This paper applies the defined framework to two real event logs.

1 Introduction

Event logs are data sets currently produced by several information systems (e.g. workflow management systems). They contain the executions (called traces) of a business process. A trace is defined as an ordered list of activities invoked by a resource from the beginning of its execution to the end. Process mining refers to the discovery, conformance and enhancement of process models from event logs. By tightly coupling event logs and process models, process mining makes possible to detect deviations, predict delays, support decision making and recommend process redesigns.

Thus far, the majority of the focus of process mining research has been on control flow, i.e. the ordering of activities. However, as discussed in [13], process mining can go beyond the control flow perspective. In particular, the organizational perspective can be considered by focusing on the resources, i.e. which performers are involved in the process model and the way they are related.

Organizational mining focuses on the organization perspective by learning more about people, machines, organizational roles, work distribution and work

© Springer International Publishing Switzerland 2016
M. Ceci et al. (Eds.): NFMCP 2015, LNAI 9607, pp. 46–60, 2016.
DOI: 10.1007/978-3-319-39315-5_4

patterns [11]. Song and van der Aalst [11], seminally, introduced the social network analysis as a comprehensive approach towards organizational mining. The nodes in a social resource network correspond to organizational entities which are, in general, one-to-one associated with resources. The arcs in a social resource network correspond to relationships between such organizational entities. Various ways have been developed in [14], in order to construct social resource networks from event logs. Quantifying the similarity of two resources is just one of many ways of constructing a social network. Every resource can be associated with a profile, i.e. a vector that describes the relevant features of a resource, while the distance between two profiles can be quantified by using well-known distance measures. A few studies [14,15] investigated how social networks analysis can be tailored for organizational mining. They mainly base on converting an event log into a social network of resources and generating social network metrics, in order to determine relevant organizational patterns. For example, between (a ratio based on the number of geodesic paths visiting a given node) can be used to find possible bottlenecks.

On the other hand, the discovery of organizational patterns cannot neglect that one of the most relevant features of social networks is the community structure of the network [9], i.e. the organization of nodes in communities, with many arcs joining nodes of the same cluster and comparatively few arcs joining nodes of different communities. Such communities can be considered as fairly independent compartments (namely organizational structure) of a network, playing a similar role. Based upon the idea that identifying communities in a network technically is finding node clusters in graphs, Song and van der Aalst [11] applied various clustering algorithms, in order to discover similar resources grouped in the same organizational structure. However, network data pose specific challenges to classical clustering algorithms [6] that we cannot neglect when looking for organizational patterns in process mining. Traditional clustering works on the distance or similarity matrix, but network data structure leads to specific algorithms using the graph property directly (k-clique, quasi-clique, vertex-betweenness, edge-betweenness).

Several community detection algorithms, exploiting graph information, have been investigated in social network analysis (see [5,9] for recent surveys). However, to the best of our knowledge, performances of these algorithms are still unexplored in the organization perspective of the process mining scenario. A further limit of the seminal research of Song and van der Aalst [11] is that they used clustering to detect communities of resources in an event log, which was processed as a static, finite dataset. However, this static view can be restrictive, as it neglects the real case of an event log that is continuously fed with new events generated from (new) running traces. On the other hand, a dynamic event log is expected to feed a dynamic social resource network, which may change over time (new resources are active, old resources are inactive). This poses the process mining problem of discovering and tracking changes in resource communities (organization structures) of a business process. This problem is also consistent with a recent trend of research [4,8,12] in social network analysis, which has started to consider the problem of tracking the progress of communities over time in a dynamic social network.

In this paper, we formalize an event-based model of the log that is handled as a stream of events. We address the task of discovering and tracking time evolving resource communities in process mining. The stream is produced by several running traces of a specific business process. Every event in the stream is time stamped, belongs to one running trace, is performed by a specific resource and executes a certain activity. A time-based window model is used to decompose the stream into consecutive windows. We formulate a two-stepped stream learning framework, named TOSTracker (Time-evolving Organizational Structure Tracker), in order to perform organizational mining of streamed resources of a business process. In the on-line phase, event data produced by running traces are queried, in order to extract a social resource network, while a community detection algorithm (i.e. Louvain algorithm [1]) is applied, in order to determine communities covering specific organizational roles. It is noteworthy that, in order to detect "overlapping" communities, we investigate the idea representing the arcs of the social resource network as nodes of a linear network [2] and apply Louvain algorithm to this linear network. In the off-line phase, the evolution (e.g. birth, death, merge, split, contraction and expansion) of discovered time-evolving communities is tracked over time.

The paper is organized as follows. In Sect. 2, we report basic concepts of this study, while in Sect. 3 we describe the organizational mining framework. In Sect. 4, we analyze the performances of this framework in a case study with real data. Finally, conclusions are drawn and future works are sketched.

2   Basics

The premise is that an *event log* $\mathcal{L}$ is a set of events concerning a certain business process type $\mathcal{P}$. An *event* $\epsilon(tid, a, r, t)$ is characterized by a set of mandatory characteristics, that is, the event corresponds to an activity a, is triggered by a resource r and has a timestamp t that represents date and time of occurrence. In addition, each event in the log is linked to a particular trace tid and is globally unique. A *trace* $\mathcal{T}$ represents the execution of a business process instance. It is a finite sequence of distinct events, that is,

$$\mathcal{T} = \epsilon(tid, a_1, r_1, t_1), \epsilon(tid, a_2, r_2, t_2), \ldots, \epsilon(tid, a_n, r_n, t_n), \tag{1}$$

such that all events of $\epsilon(tid, a_i, r_i, t_i) \in \mathcal{T}$ are linked to a specific trace tid, while time is non-decreasing in the trace (i.e. for $1 \leq i < j \leq n : t_i \leq t_j$). According to this definition of trace, an event log is traditionally dealt as a static bag of full traces of a specific business process [13]. An example showing a fragment of an event log, organized around the concept of trace, is reported in Table 1.

In this paper, we move from a static perspective to a dynamic perspective when handling event logs. We consider an event log as a dynamic dataset that is continuously fed with new events generated by traces, which are still running. This allows us to handle an event log according to an *event stream* model. In particular, an event stream $\mathcal{S}$ is an ordered, unbounded sequence of time stamped events:

$$\mathcal{S} = \epsilon(tid_1, a_1, r_1, t_1), \epsilon(tid_2, a_2, r_2, t_2), \ldots, \epsilon(tid_i, a_i, r_i, t_i), \ldots \tag{2}$$

Table 1. A fragment of an example event log. Each event is linked to a specific trace. It corresponds to an activity, has a timestamp and is triggered by a resource.

TraceId	Activity	Timestamp	Resource
1	Register request (R)	2010-12-30:11:02	Pete
1	Examine throughly (ET)	2010-12-31:10:06	Sue
1	Check ticket (CT)	2011-01-05:15:12	Mike
1	Decide (D)	2011-01-06:11:18	Sara
1	Reject request (RR)	2011-01-07:14:24	Pete
2	Register request (R)	2010-12-30:11:32	Mike
2	Check ticket (CT)	2010-12-30:12:12	Mike
2	Examine causally (EC)	2010-12-30:14:16	Pete
2	Check ticket (CT)	2010-12-31:15:31	Pete
2	Decide (D)	2011-01-05:11:22	Sara
2	Pay compensation (PC)	2011-01-08:12:05	Ellen
3	Register request (R)	2010-12-30:14:32	Pete
3	...	...	...
...	...	...	...

where events of the stream arrive sequentially, at consecutive time points (i.e. $t_i \leq t_{i+1}$), from the bag of running traces of business process P. The bag of running traces may change over time, since old executions can be completed, while new executions can be started at a certain time point. The event stream associated with the fragment of event log reported in Table 1 is shown in Table 2.

An event stream, like any data stream, is unbounded in length. It is impractical to query all the data of a stream. Windows are commonly used stream approaches to query open-ended data. Instead of computing an answer over the whole data stream, the query (or operator) is computed, maybe several times, over a finite subset of events. Several window models are defined in the literature. In this study, we consider the *time-based window model* [3], which decomposes an event stream into consecutive (non overlapping) windows of fixed temporal size. When a window is completed, it is queried. The answer is stored in a data synopsis for the mining phase, while the windowed data are discarded. Formally, let $\Delta(T)$ be the window temporal size of the model, a time-based window model decomposes a stream S into non overlapping windows,

$$S(\Delta(T)) = t \to t + \Delta(T), t + \Delta(T) \to t + 2\Delta(T), \ldots, t + (j-1)\Delta(T) \to t + j\Delta(T), \ldots \quad (3)$$

so that every window $t + (j-1)\Delta(T) \to t + j\Delta(T)$ selects stream events $\epsilon(tid_i, a_i, t_i, r_i) \in S$ acquired at each time point $t \in [t + (j-1)\Delta(T), t + j\Delta(T)[$.

As a query operator, we consider a *social resource network* constructor. Consequently, as a data synopsis for the data storage, we use a *weighted graph*. The nodes of this social network (or equivalently graph data synopsis) correspond to resources triggering one or more events in the data window. Each resource is

Table 2. The stream model of the event log reported in Table 1. The event stream is divided into consecutive windows, which are 24 h long according to the time-based window model with $\Delta(t) = 24$ h.

TraceId	Activity	Timestamp	Resource
1	Register request (R)	2010-12-30:11:02	Pete
2	Register request (R)	2010-12-30:11:32	Mike
2	Check ticket (CT)	2010-12-30:12:12	Mike
2	Examine causally (EC)	2010-12-30:14:16	Pete
3	Register request (R)	2010-12-30:14:32	Pete
2	Check ticket (CT)	2010-12-31:15:31	Pete
1	Examine throughly (ET)	2010-12-31:10:06	Sue
2	Decide (D)	2011-01-05:11:22	Sara
1	Check ticket (CT)	2011-01-05:15:12	Mike
1	Decide (D)	2011-01-06:11:18	Sara
1	Reject request (RR)	2011-01-07:14:24	Pete
2	Pay compensation (PC)	2011-01-08:12:05	Ellen
...	...	...	...

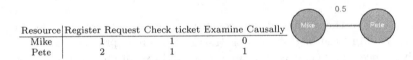

Resource	Register Request	Check ticket	Examine Causally
Mike	1	1	0
Pete	2	1	1

Fig. 1. The activity resource profile of Pete and Mike (left side). Both are resources active in the data window [2010-12-30:0:00, 2010-12-30:24:00] of the stream reported in Table 2. The social resource network extracted from this data window (right side) contains two nodes associated to resources Pete and Mike, as well as an arc connecting these resources. The weight associated to the arc is the Pearson correlation coefficient computed between the activity resource profiles of the edged nodes.

associated with a *resource-activity profile* that represents the number of times a resource performs an activity in the data window. Arcs between nodes are associated with weights that express the importance of the relations. As in [11], for each pair of resources r_i and r_j, we compute the Pearson correlation coefficient of the resource-activity profiles, which are associated with r_i and r_j, respectively.

Formally, $w(r_i, r_j) = \dfrac{\sum_A r_i(A)r_j(A) - n\overline{r_i}\,\overline{r_j}}{\sqrt{\sum_A r_i(A)^2 - n\overline{r_i}^2}\sqrt{\sum_A r_j(A)^2 - n\overline{r_j}^2}}$ where A denotes

an activity associated with the resource profiles, $r_i(A)$ $(r_j(A))$ is the number of times A is performed by r_i (r_j) in the data window, $\overline{r_i(A)}$ $(\overline{r_j(A)})$ is the average $\overline{r_i(A)} = \frac{1}{n}\sum_A r_i(A)$ $(\overline{r_j(A)} = \frac{1}{n}\sum_A r_j(A))$ and n is the number of activities in

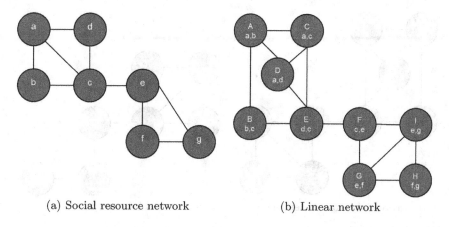

(a) Social resource network   (b) Linear network

Fig. 2. Overlapping resource community discovery: the social resource network (see (a)) is transformed into the linear network (see (b)).

the resource profile. We rank potential arcs according to the Pearson correlation values associated with. Arcs associated with the top p ranked Pearson correlation are, finally, added to the graph data synopsis. An example of a social resource network is reported in Fig. 1. Alternative metrics, which can be computed to estimate the weight of an arc, can monitor handover of work or subcontracting [11].

3   Time-Evolving Organization Structure Tracker

The organizational mining framework, called TOSTracker, operates in two phases. The on-line phase consumes events as they arrive from the event stream and analyzes the buffered events, window-by-window, in order to determine a social resource network, which is stored in a graph data synopsis. For each window, (overlapping) communities are detected from the social resource network stored in the graph-based data synopsis (see details in Sect. 2). The set of resource communities is then discovered from this data synopsis as a model of the organization structure of the business process along the time horizon associated to the processed event window. Since this resource community model is stored in a database, while windowed events are definitely discarded, resource communities represent the organizational knowledge for the future off-line query phase. The off-line phase, which is repeatable, tracks the evolution of resource communities discovered along a query time horizon and retrieved from the database.

3.1   Resource Community Detection

Resource community detection is performed by resorting to Louvain algorithm [1]. This is a greedy optimization that attempts to optimize the "modularity" of a partition of the network.

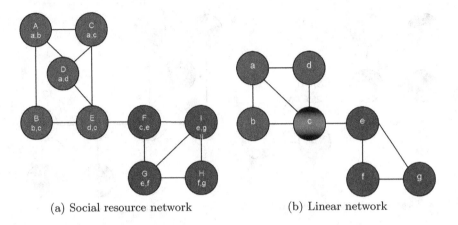

(a) Social resource network   (b) Linear network

Fig. 3. Overlapping resource community discovery: Disjoint communities discovered by Louvain algorithm in the linear network (see (a)) are mapped into overlapping communities of resources (see (b)). We note that node c of the social resource network belongs to the red community with degree 0.75 and to the blue community with the degree 0.25 (Color figure online).

The modularity is a measure of the structure of a network. It is designed to measure the strength of division of a network into communities (or clusters). Formally, modularity is the fraction of the arcs that fall within the given communities minus the expected such fraction if edges were distributed at random. So, for a given division of the network's nodes into some communities, modularity reflects the concentration of arcs within modules compared with random distribution of arcs between all nodes regardless of communities. Networks with high modularity have dense connections between the nodes within communities, but sparse connections between nodes in different communities. In this study, the randomization of the arcs is done according to the Configuration model presented in [7], as it allows us to generate a randomization preserving the node degrees of the original network. The modularity is computed according to the measure of Reichardt-Bornholdt [10].

Let $\mathcal{G}(\mathcal{N}, \mathcal{A})$ be a network, N be the set of nodes and A be the set of arcs. The optimization of modularity is performed in two steps. In the first step, Louvain algorithm looks for "small" communities by optimizing modularity locally. So it, initially, assigns a different community to each node $u \in \mathcal{N}$. Hence, in this initial partition, there are as many communities as there are nodes. Then, for each node $u \in \mathcal{N}$, Louvain algorithm considers neighbors v of u (i.e. each v is edged to u in $\mathcal{A}$) and evaluates the gain of modularity that would take place in the network by transferring u from its community to the community of v. The node u is, finally, transferred to the community for which this gain is positive and maximum. If no positive gain is possible, it stays in its original community. This process is applied repeatedly and sequentially for all nodes until no further improvement can be achieved and the first phase is then complete. In the second step, Louvain algorithm aggregates nodes belonging to the same community and builds a new network whose nodes are the communities. This step of the algorithm consists in building a new network whose nodes are now the communities found during the first phase. The weights of the arc edging two community nodes is given by the sum of the weight of the arcs between nodes in the corresponding two communities. Once this new network is computed, the first step of the algorithm is applied to the resulting weighted network. In this way, the two steps are repeated iteratively until a maximum of modularity is attained and a hierarchy of communities is produced.

We note that Louvain algorithm is an efficient and easy-to-implement algorithm for identifying communities in networks. However, it allows us to detect only "disjoint" communities. In contrast, organizational structures of a business process may be overlapping in a social resource network (i.e. a resource may place two roles in the same business process). In order to discover overlapping resource communities, we have applied Louvain algorithm to the linear network that is constructed from the social resource network. This linear network is constructed by representing the arcs of the social resource network as nodes of the linear network [2]. Let u and v be two nodes of the linear network, so that u denotes the arc (u_i, u_j, w_u) of the social resource network, while v denotes the arc (v_i, v_j, w_v) of the social resource network. There is an arc (u, v, w) in the linear network iff arcs (u_i, u_j, w_u) and (v_i, v_j, w_v) share one vertex in the social resource network (i.e. $u_i = v_i$ or $u_i = v_j$ or $u_j = v_i$ or $u_j = v_j$). Let x be the vertex shared between (u_i, u_j, w_u) and (v_i, v_j, w_v), weight $w = w_u/(deg(x) - w_v)$ where the $deg(x)$ is the sum of weights associated to arcs incoming/outcoming x in the social resource network (see Fig. 2(a)–(b)).

Disjoint communities discovered in the linear network are then mapped into possibly overlapping communities discovered in the social resource network. A node u_i of the social resource network belongs to every community discovered in the linear network, which groups at least one node u of the linear network, so that u is associated it an arc of the social resource network with a vertex in u_i (see Fig. 3(a)–(b)). Let u_i a node of the social resource network, the degree according to the resource u_i belongs to a community C can be computed as follows:

$$degree(u_i, C) = \frac{|\{C' \in C_{u_i} | C' = C\}|}{|\{C_{u_i}\}|} \qquad (4)$$

where C_{u_i} is the set of communities assigned to arcs of I_{u_i} in the linear network, while I_{u_i} is the set of arcs of the social resource network incoming/outcoming u_i (see Fig. 3(b)).

3.2 Tracking Evolutions of Resource Communities

A dynamic resource community is represented as a time line, that is, a sequence of (evolving) resource communities, ordered by time, with at least one resource

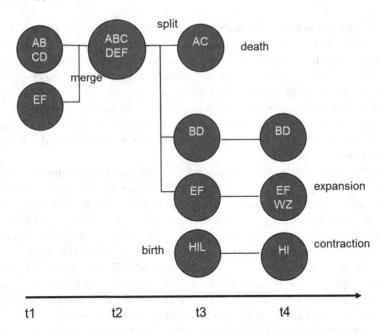

Fig. 4. The time line of the evolutions of dynamic resource communities

community for each time point t. The evolutions of dynamic resource communities along their time line is expressed in terms of birth, death, merge, split, expansion and contraction [4] (see Fig. 3). Intuitively, the *birth* event describes a new resource community observed at time t with no corresponding resource community in the set of communities already tracked. A new community is created with time line starting at the birth time. The *death* event describes the dissolution of a resource community that does not appear for several consecutive time points. The timeline of this community ends when it disappears. The *merge* event occurs when two or more distinct resource communities observed at time $t - 1$ can be similar to a single community at time t. A branch is added to connect the single communities at time $t - 1$ to the merged community created at time t. The single communities, subsequently, share the same time line. The *split* event occurs when a single resource community present at time $t - 1$ can be similar to two or more distinct resource communities at time t. A branching occurs from the starting community to the split ones with the creation of an additional resource community that shares the timeline of up to time $t - 1$, but has a distinct timeline from time t onwards. The *expansion* of a resource community occurs when its cardinality grows at time t. The *contraction* of a resource community occurs when its cardinality decreases at time t. The Jaccard coefficient can be computed, in order to estimate the similarity between communities and detect these events.

Algorithm 1. Resource community tracking

Require: Stream of resource communities {initialization phase $(t = 1)$}
1: $\mathcal{C}_1 \leftarrow$ communitySet(1)
2: $\mathcal{F} \leftarrow \mathcal{C}_1$ {Mining phase $(t = i)$}
3: $\mathcal{C}_i \leftarrow$ communitySet(t)
4: **for** $(C, C_F) \in \mathcal{C}_i \times \mathcal{F}$ **do**
5:     $j \leftarrow$ jaccard(c, f)
6:     Update $\mathcal{F}$ according to events detected processing j
7: **end for**

The evolution is tracked according to the algorithm described in [4]. This algorithm is independent on the algorithm selected to discover online the resource communities. It is based on the analysis of the degree of similarity between pairs of communities discovered at consecutive time points. The input is a time series $\mathcal{T}$ of resource community sets (i.e. $\mathcal{C}_1, \mathcal{C}_2, \ldots, \mathcal{C}_t, \ldots, \mathcal{C}_n$), where each set $\mathcal{C}_t$ represents the organizational structure (i.e. set of resource communities) extracted over a specific time horizon and time stamped with t in $\mathcal{T}$.

The tracking algorithm is two stepped (see Algorithm 1). In the initialization phase, the algorithm constructs a dynamic resource community for each resource community $C \in \mathcal{C}_1$. The same community is also added to the set of frontier communities $\mathcal{F}$. In the mining step, the algorithm iteratively analyzes each organization structure $\mathcal{C}_t$ (with $t = 2, 3, n$) by computing the similarity between communities of $\mathcal{C}_t$ and frontier communities of $\mathcal{F}$.

Formally, for every $C \in \mathcal{C}_t$ and $C_\mathcal{F} \in \mathcal{F}$, $jaccard(C, C_\mathcal{F})$ is computed with $jaccard(\cdot, \cdot)$ the Jaccard coefficient. The compared communities are similar if and only if their Jaccard coefficient exceeds a user-defined threshold σ. We note that the output of the similarity computation may reveal series of community evolution events. These events are detected, in order to update accordingly the time lines of the dynamic resource communities, as well as the set of frontier communities. Let us consider $C \in \mathcal{C}_t$. If there is no frontier resource community $C_\mathcal{F} \in \mathcal{F}$ so that $jaccard(C, C_\mathcal{F}) \geq \sigma$ then a birth event is happened. A new dynamic community containing C is created. If there is one and only one frontier resource community $C_\mathcal{F} \in \mathcal{F}$ so that $jaccard(C, C_\mathcal{F}) \geq \sigma$, this indicates the continuation of C. If there are two or more frontier resource community $C_{\mathcal{F}_1}, \ldots, C_{\mathcal{F}_k} \in \mathcal{F}$ so that $jaccard(C, C_{\mathcal{F}_i}) \geq \sigma$ $(i = 1, 2, \ldots, k)$, then a merge event is happened. The new merged community is connected to the time lines of the frontier communities contributing to the merge. On the other hand, let us consider $C_\mathcal{F} \in \mathcal{F}$. If there are two or more resource community $C_1, \ldots, C_k \in \mathcal{C}_t$ so that $jaccard(C_i, C_\mathcal{F}) \geq \sigma$ $(i = 1, 2, \ldots, k)$, then a split event is happened. Every new split community is connected to the time line of the frontier community that has originated the split. Finally let us consider the pair $(C, C_\mathcal{F}) \in \mathcal{C}_t \times \mathcal{F}$ so that $jaccard(C, C_\mathcal{F}) \geq \sigma$ then if $cardinality(C - C_\mathcal{F}) > 0$, then an expansion event is happened; if $cardinality(C_\mathcal{F} - C) > 0$, then a contraction event is manifested.

4   Case Studies

We have evaluated performances of the proposed framework by considering real traces of two distinct business processes. The former business process is taken from a Dutch Financial Institute and provided in the Business Processing Intelligence Challenge 2012 (BPI 2012).[1] The log contains 262.200 events from September 27, 2011 to March 10, 2012, 69 resources, in 13.087 traces. The business process is an application process for a personal loan or overdraft within a global financing organization. The event stream has been processed with time-based window model with $\Delta(T) = 15$ days. The latter business process is taken from a Dutch Academic Hospital and provided in the Business Processing

Table 3. Extracted social resource networks (BPI 2012 and BPI 2013): number of nodes and number of arcs.

time horizon	n. nodes (active resources)	n. arcs
BPI 2012		
2011/09/27 00:00:00-2011/10/12 00:00:00	46	486
2011/10/12 00:00:00-2011/10/27 00:00:00	45	419
2011/10/27 00:00:00-2011/11/10 00:00:00	46	447
2011/11/10 00:00:00-2011/11/25 00:00:00	50	499
2011/11/25 00:00:00-2011/12/10 00:00:00	50	547
2011/12/10 00:00:00-2011/12/25 00:00:00	51	582
2011/12/25 00:00:00-2012/01/09 00:00:00	47	538
2012/01/09 00:00:00-2012/01/24 00:00:00	45	507
2012/01/24 00:00:00-2012/02/08 00:00:00	47	566
2012/02/08 00:00:00-2012/02/23 00:00:00	48	605
2012/02/23 00:00:00-2012/03/09 00:00:00	46	571
2012/03/09 00:00:00-2012/03/24 0:00:00	40	411
BPI 2011		
2005/01/01 00:00:00-2005/05/01 00:00:00	22	13
2005/05/01 00:00:00-2005/08/29 00:00:00	19	8
2005/08/29 00:00:00-2005/12/27 00:00:00	27	24
2005/12/27 00:00:00-2006/04/26 00:00:00	29	23
2006/04/26 00:00:00-2006/08/24 00:00:00	28	21
2006/08/24 00:00:00-2006/12/22 00:00:00	29	16
2006/12/22 00:00:00-2007/04/21 00:00:00	27	20
2007/04/21 00:00:00-2007/08/19 00:00:00	27	24
2007/08/19 00:00:00-2007/12/17 00:00:00	27	24
2007/12/17 00:00:00-2008/03/20 00:00:00	19	9

---

[1] http://www.win.tue.nl/bpi/2012/challenge.

Intelligence Challenge 2011 (BPI 2011).[2] The log contains 150.291 events form January 3, 2005 to March 20, 2008 in 1143 traces. Each trace is a patient of a Gynaecology department. It indicates how the patient goes through different (maybe overlapping) phases, where a phase consists of the combination Diagnosis and Treatment. The event stream has been processed with time-based window model with $\Delta(T) = 15$ days. In both business processes, the resource social network of every window is buffered into a graph data synopsis, where nodes are associated with resources triggering an event in the window, arcs between nodes are materialized if the weight associated with the arc is in the top $p = 75\%$ Pearson correlation coefficients computed between the activity-resource profiles of nodes (see details in Sect. 2). A summary of property of extracted social resource networks, extracted from both business processes, is reported in Table 3.

The overlapping resource communities of the organizational structure are discovered in the data synopsis where event data are buffered, according

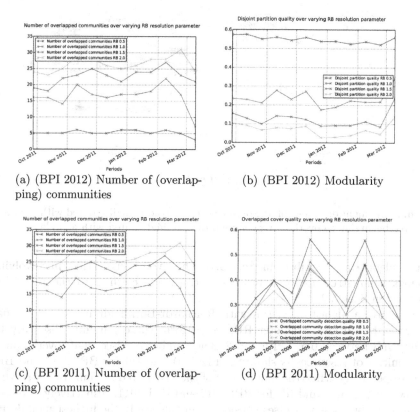

(a) (BPI 2012) Number of (overlapping) communities

(b) (BPI 2012) Modularity

(c) (BPI 2011) Number of (overlapping) communities

(d) (BPI 2011) Modularity

Fig. 5. Organizational structure discovery by varying the resolution parameter RB of Reichardt-Bornholdt measure between 0.5, 1, 1.5 and 2.

---

[2] http://www.win.tue.nl/bpi/2011/challenge.

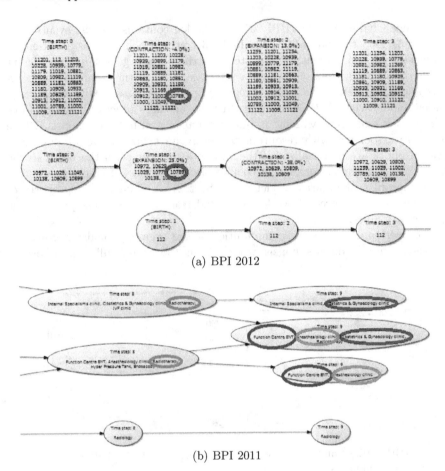

(a) BPI 2012

(b) BPI 2011

Fig. 6. A fragment of time-evolving organizational structures discovered with RB=0.5. Resource belonging to overlapping communities are highlighted with a circle.

to the algorithm described in Sect. 3.1. The evolution of the time-evolving organizational structure of a process is tracked according to the algorithm described in Sect. 3.2.

As Louvain algorithm is run with Reichardt-Bornholdt measure [10] that requires a resolution parameter RB, we perform the community discovery by varying RB between 0.5, 1, 1.5 and 2. The number of discovered (overlapping) communities of resources is shown in Fig. 5(a) and (c) for BPI 2012 and BPI 2011, respectively, while the modularity of the discovered organization structure is shown in Fig. 5(b) and (d) for BPI 2012 and BPI 2011, respectively. We note that in both business processes considered in this study the highest modularity is achieved when the lowest number of communities is detected per window in correspondence of $RB = 0.5$.

Finally, we track the evolution of the organizational structure discovered with $RB = 0.5$. An example of the evolutions tracked is shown in Fig. 6(a) for BPI

2012 and in Fig. 6(b) for BPI 2011. In both cases, we are able to see the evolutions (birth, split, merge, and so on) of the resource communities discovered by analyzing the events produced by the executions of the two business processes considered in this study. It is noteworthy that, in both examples, several resources participate to different organizational structures simultaneously. This phenomenon is appropriately captured by the discovery of overlapping communities.

5   Conclusions

In this paper, we have described a framework to perform organizational mining of streamed resources of a business process. The stream is processed according to a time-based window model. Events batched in a window are queried, in order to extract a social resource network. A community detection algorithm is applied, in order to determine the organization structure of the business process, i.e. possibly (overlapping) communities of resources covering specific organizational roles. The evolution (e.g. birth, death, merge, split, contraction and expansion) of the discovered communities is tracked over time. In this study, social resource network is represented as undirected graph. The effectiveness of the proposed studies is investigated in two case studies As future work, we plan to extend this framework to process directed social resource networks. In addition, we plan to investigate the opportunity of performing, incrementally, the discovery of communities by integrating the tracking phase in the community discovery phase.

Acknowledgments. This work fulfills the research objectives of the PON 02_00563_3470993 project "VINCENTE - A Virtual collective INtelligenCe ENvironment to develop sustainable Technology Entrepreneurship ecosystems", funded by the Italian Ministry of University and Research (MIUR), as well as the project "LOGIN-LOGistica INTegrata 2012–2015 (PII INDUSTRY 2015)", announcement New Technologies for the Made in Italy.

References

1. Blondel, V., Guillaume, J.-L., Lambiotte, R., Lefebvre, E.: Fast unfolding of communities in large networks. J. Stat. Mech.: Theory Exp. **10**, P10008 (2008)
2. Evans, T., Lambiotte, R.: Line graphs of weighted networks for overlapping communities. Eur. Phys. J. B **77**(2), 265–272 (2010)
3. Gaber, M.M., Zaslavsky, A., Krishnaswamy, S.: Mining data streams: a review. ACM SIGMOD Rec. **34**(2), 18–26 (2005)
4. Greene, D., Doyle, D., Cunningham, P.: Tracking the evolution of communities in dynamic social networks. In: ASONAM 2010, pp. 176–183 (2010)
5. Harenberg, S., Bello, G., Gjeltema, L., Ranshous, S., Harlalka, J., Seay, R., Padmanabhan, K., Samatova, N.: Community detection in large-scale networks: a survey and empirical evaluation. Wiley Interdisc. Rev.: Comput. Stat. **6**(6), 426–439 (2014)

6. Lei, T., Huan, L.: Community Detection and Mining in Social Media. Morgan and Claypool Publishers, San Rafael (2010)
7. Newman, M.E.J.: Finding community structure in networks using the eigenvectors of matrices. Phys. Rev. E **74**(3), 036–104 (2006)
8. Oliveira, M.D.B., Guerreiro, A., Gama, J.: Dynamic communities in evolving customer networks: an analysis using landmark and sliding windows. Soc. Netw. Analys. Min. **4**(1), 208 (2014)
9. Plantie, M., Crampes, M.: Survey on social community detection. In: Ramzan, N., van Zwol, R., Lee, J.-S., Clüver, K., Hua, X.-S. (eds.) Social Media Retrieval. Computer Communications and Networks, pp. 65–85. Springer, London (2013)
10. Reichardt, J., Bornholdt, S.: Statistical mechanics of community detection. Phys. Rev. E **74**(1), 016110 (2006)
11. Song, M., van der Aalst, W.M.P.: Towards comprehensive support for organizational mining. Decis. Support Syst. **46**(1), 300–317 (2008)
12. Spiliopoulou, M.: Evolution in social networks: A survey. In: Aggarwal, C.C. (ed.) Social Network Data Analytics, pp. 149–175. Springer US, New York (2011)
13. van der Aalst, W.M.P.: Process Mining - Discovery Conformance and Enhancement of Business Processes. Springer, Heidelberg (2011)
14. van der Aalst, W.M.P., Reijers, H.A., Song, M.: Discovering social networks from event logs. Comput. Support. Coop. Work **14**(6), 549–593 (2005)
15. van der Aalst, W.M.P., Song, M.: Mining social networks: uncovering interaction patterns in business processes. In: Desel, J., Pernici, B., Weske, M. (eds.) BPM 2004. LNCS, vol. 3080, pp. 244–260. Springer, Heidelberg (2004)

Intelligent Adaptive Ensembles for Data Stream Mining: A High Return on Investment Approach

M. Kehinde Olorunnimbe[1], Herna L. Viktor[1(✉)], and Eric Paquet[1,2]

[1] School of Electrical Engineering and Computer Science, University of Ottawa,
Ottawa, ON K1N 6N5, Canada
{molor068,hviktor}@uottawa.ca
[2] National Research Council of Canada, Ottawa, ON K1A 0R6, Canada
eric.paquet@nrc-cnrc.gc.ca

Abstract. Online ensemble methods have been very successful to create accurate models against data streams that are susceptible to concept drift. The success of data stream mining has allowed diverse users to analyse their data in multiple domains, ranging from monitoring stock markets to analysing network traffic and exploring ATM transactions. Increasingly, data stream mining applications are running on mobile devices, utilizing the variety of data generated by sensors and network technologies. Subsequently, there has been a surge in interest in mobile (or so-called pocket) data stream mining, aiming to construct near real-time models. However, it follows that the computational resources are limited and that there is a need to adapt analytics to map the resource usage requirements. In this context, the resultant models produced by such algorithms should thus not only be highly accurate and be able to swiftly adapt to changes. Rather, the data mining techniques should also be fast, scalable, and efficient in terms of resource allocation. It then becomes important to consider Return on Investment (ROI) issues such as storage space needs and memory utilization. This paper introduces the Adaptive Ensemble Size (AES) algorithm, an extension of the Online Bagging method, to address this issue. Our AES method dynamically adapts the sizes of ensembles, based on the most recent memory usage requirements. Our results when comparing our AES algorithm with the state-of-the-art indicate that we are able to obtain a high Return on Investment (ROI) without compromising on the accuracy of the results.

Keywords: Data streams · Metalearning · Adaptive ensemble size · Return on investment · OzaBag

1 Introduction

In this era of the Internet of Things and Big Data, a proliferation of connected devices continuously produce massive amounts of fast evolving streaming data. A number of online learning methods have been highly successful in constructing accurate models against massive data streams. Success stories include retail store

© Springer International Publishing Switzerland 2016
M. Ceci et al. (Eds.): NFMCP 2015, LNAI 9607, pp. 61–75, 2016.
DOI: 10.1007/978-3-319-39315-5_5

sales streamlining, chemical plant shutdown time prediction, and stock market monitoring, amongst others [10, 15]. Currently, the development of techniques to facilitate mobile (or pocket) data stream mining is an emerging area of research with applications in the areas of business, telemedicine and security, amongst others. In this setting, the resultant models produced by data stream mining algorithms should not only be highly accurate and be able to swiftly adapt to changes. Rather, the learning techniques should also be efficient in terms of resource allocation. It then becomes important to consider issues such as storage space needs and memory utilization. This is especially relevant when we aim to build personalized, near-instant models for Big Data on small devices [10, 14, 16].

This research addresses this emerging need for accurate, yet efficient, model construction [1, 20]. Our aim is to take an adaptive approach to resource allocation during the mining process. Consideration is given to the memory available to the algorithm and the speed at which data is processed. To this end, we introduce the Adaptive Ensemble Size (AES) technique that extends the Online Bagging (OzaBag) online ensemble learning algorithm. Our AES method takes advantage of the memory utilization cost, in order to vary the ensemble size during the data mining process. We aim to minimize the memory usage, while maintaining highly accurate models with a high utility. The reasoning behind our approach is based on the following observation. Intuitively, a higher change in memory utilization during stream classification potentially indicates a shift in the data at that particular stream window, which may occur as a result of concept drift. That is, the learners receive new instances that are potentially more difficult to classify, thus leading to the utilization of more memory in order to expand the current models, in an attempt to maintain high accuracy. In this case, our AES algorithm increases the ensemble size in an attempt to maintain higher accuracies. The inverse is also true, in that we may be able to reduce the ensemble size, with no or little cost in terms of accuracies, when the properties of the data remain stable.

This paper is organized as follows. Section 2 introduces background work. In Sect. 3, we detail the Adaptive Ensemble Size (AES) algorithm. Section 4 discusses our experimental evaluation and results. Finally, Sect. 5 presents our conclusion and highlights future work.

2   Background

In online learning, an algorithm learns one instance at a time, on arrival, without the need for storage and reprocessing, and the models built reflects the data seen so far [18]. This characteristic makes it a highly suitable approach for use in a streaming environment where there is a requirement to construct models from partial datasets and where the datasets are often too large to maintain on secondary storage. Oza and Russell's Online Bootstrap Aggregating (OzaBag) is an online version of the well-known Bagging ensemble learning algorithm [5, 18, 19]. The key idea behind the Bagging (or Bootstrap Aggregation) approach, as originally created by Breiman, is that the members of an ensemble vote with equal weight [5].

Using a number of classifiers (e.g. decision trees) as base learners, and a dataset N, Bagging draws samples with replacement, and induces models from each of the learners independently. For a new classification problem, the most predicted class amongst the models is assigned to the new instance. This value is obtained by a combined vote of all the classes within the different models. Bagging works best with an unstable base learner, because each independently created model will be very different in response to minor changes [3]. This makes it particularly useful in a data stream with concept drift, because there is a very high chance that the samples the models are built on are different. Bagging thus aid to reduce the bias and variance by averaging across all models into a single result.

For a dataset N, the Bagging algorithm assumes a binomial distribution of the data, due to the finite size of the dataset in memory. OzaBag, on the other hand, assumes a Poisson distribution because of the continuous nature $(N \rightarrow \infty)$ of the stream. Hence, the number of copies are determined using the Poisson distribution. Similar to Bagging algorithm, the most predicted class amongst the models is assigned to a new instance of the classification problem.

OzaBag with ADWIN is an extension of the OzaBag algorithm, aiming to facilitate concept drift by increasing the rate of resampling [2]. Specifically, ADWIN is a change detector that was implemented as a way to estimate changes in a data stream with concept drift using an adaptive sliding window. In this approach, whenever the difference between the mean values of the elements in two sliding windows is more than a threshold value, the older sliding window is dropped. The window size(s) are not maintained explicitly, as ADWIN uses a variant of exponential histogram technique as shown in Fig. 1. It keeps an approximate number of $1's$ as the bucket size in a sliding window W using $O(M \cdot log(W/M))$ memory and $O(logW)$ processing time. Thus, ADWIN is in essence an estimator that keeps a variable-length window of recently seen items. The maximal length of the window must be statistically consistent with the following hypothesis: there has been no change in the average value inside the window [8]. In Ozabag with ADWIN, a large Poisson distribution parameter (λ) is used to increase the resampling weight. While the optimal value of λ may not be the same for different datasets, by increasing its value, the diversity of the weight is increased. This effectively modifies the input space of the

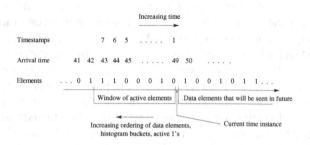

Fig. 1. Exponential histogram illustration [7]

classifiers inside the ensemble. Interested readers are referred to [14] for a detailed discussion of ADWIN.

The two variants of the OzaBag algorithm are available in the Massive On-line Analysis (MOA) data stream mining framework [4]. In our research, we extended the OzaBag algorithms and used the Hoeffding Tree method as base learner [8]. This incremental, anytime decision tree algorithm exploit the fact that a small sample may often be enough to choose an optimal splitting attribute and has been shown to have sound guarantees of performance [8].

Recall that our goal is to study the cost of building and applying data mining models while taking the memory usage of the ensemble, as well as the ensemble size, into consideration. To this end, we performed a series of preliminary experiments in order to observe the effects of concept drift on memory utilization. We also investigated the influence of ensemble sizes on the accuracy.

2.1 Size of Ensemble Object in Memory

In our work, we start from the observation that the size of an ensemble object in memory tends to be proportional to changes in characteristics of the learning instances. In this section, we present some preliminary experimentation in order to further ascertain this claim. To this end, we simulated two artificial data streams with MOA, using the LED data stream generator [4], which has been widely used in many studies [2,9]. In this dataset, the task is to predict the number displayed on a 7-segment LED display. The artificial streams are generated without noise, and we simulate concept drift as applied to a different number of its 7 attributes. This allows us to monitor the effect of concept drift with the memory consumed in the learning process. Having zero noise in the stream helps us to minimize the other factors that might have effect on memory utilization.

In first our experiments, one stream is simulated with concept drift while the other is not. In this setting, concept drift was gradually induced throughout the learning process. Figure 2, shows the memory usage and memory change as more instances are evaluated. The graphs indicate the memory usage while the learning process is being completed. As we can observe from Fig. 2a, in a steady stream with no randomness or drift, the memory usage rapidly stagnates and becomes stable, as every new ensemble presents exactly the same size than the previous one. On the other hand, in Fig. 2b, there is an increase in memory consumption in a stream with concept drift. Also, generally speaking, both OzaBag and OzaBagADWIN have a similar memory usage, for this dataset.

We conducted further experiments where we induced, and removed, concept drift at regular intervals (every 20 %) as shown in Fig. 3. Different degrees of abrupt drift were induced, ranging from a drift affecting a single attribute, 80 % drift affecting 3 attributes and a drift affecting all seven attributes. We started this experiment with 0 % drift. At 20 % stream, we introduced a slight drift, affecting 1 of the 7 attributes. We noticed a slight increase in memory usage as a result of this, which remains steady for the duration of the section. At 40 % stream, we introduced a full concept drift, affecting all 7 attributes. A much higher memory increase is observed

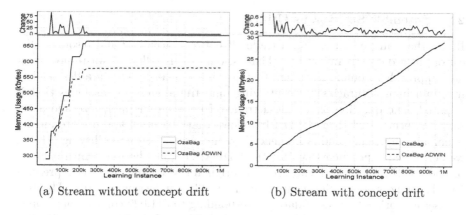

(a) Stream without concept drift   (b) Stream with concept drift

Fig. 2. Memory increase in the data stream

at this point. Our premise is that while other random factors might be able to influence a memory change in any stream, concept drift will cause a noticeable change in memory, and have a direct effect on the stream mining cost. At 60 %, the concept drift was removed. While there was a modest increase in memory consumption for that change, it gradually subsided to the level it was before the change. We introduced a drift affecting 3 attributes at 80 %, and we can see another increase in memory, relative to the concept drift. This shows that, for this artificial dataset, an increase in concept drift triggers a proportional increase in memory. In this work, we take advantage of these changes to improve the stream mining accuracy of a metalearning algorithm by way of increasing the ensemble size only when deemed beneficial.

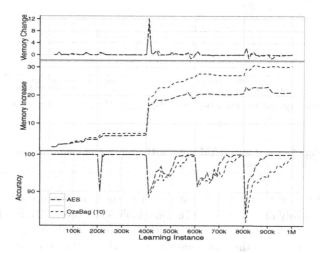

Fig. 3. Memory change with concept drift induced at regular intervals

2.2   Ensemble Size versus Utility

Recall that our goal is to find a trade-off between accuracy and resource allocation, in a data stream setting. That is, our aim is follow a 'waste not, want not' approach to online learning. To that end, our algorithms are designed to maintain high accuracies while utilizing only the amount of resources that is required. The question of finding a trade-off between ensemble size and accuracy has been studied in [11,17]. It follows that the most appropriate answer is domain dependent and is influenced by numerous factors. Following [11,17], we performed experiments against the KDD CUP 1999 dataset, which is based on the DARPA98 network capture. The original raw training dataset represents 7 weeks of network traffic, containing 4,898,431 connection records. The test dataset comprises 2 weeks of data, corresponding to 311,029 connection records. Both contains 42 features, including the label. We used the OzaBag and OzaBagADWIN algorithms, varying the ensemble sizes from 10 to 100, by intervals of 10, with Hoeffding Trees as base learners. Figure 4a, shows the error (100 %− accuracy) plotted against the ensemble size. As shown in the figure, the error rates are comparable against all ensemble sizes. That is, there is little gain in employing more base learners against this dataset. Figure 4b further shows that the computational cost of the algorithms dramatically increases when the ensemble size exceeds 20.

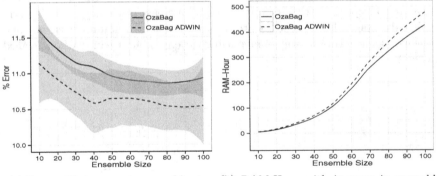

(a) Error with increase in ensemble size   (b) RAM-Hour with increase in ensemble size

Fig. 4. Error and RAM-hour with increase in ensemble size

From Table 1, we further confirm that the computational cost of the algorithms considerably increases when the ensemble size exceeds 20. For instance, if we consider the lower cost OzaBag algorithm as an example, we notice from the table that a variation of the ensemble size from 10 to 20 yields a small accuracy (k^+) improvement of 0.5474 % but at the expense of a four folds increase in terms of cost. When the ensemble size is increased to 40, the accuracy increases by a meager 0.0364 % but at the expense of a four folds increase in terms of cost as

Table 1. Error, Time and RAM-hour for different ensemble sizes

	OzaBag classifier			OzaBagADWIN classifier		
Size	% Error	Time	RAM-Hr	% Error	Time	RAM-Hr
10	11.6899	258.1505	3.7057	11.1555	289.2415	4.0624
20	11.1423	545.5511	15.8006	10.7695	630.9148	19.4642
30	11.2444	828.5681	35.5382	10.7632	931.4664	39.7808
40	11.1055	1093.4734	62.5377	10.5776	1229.6467	70.7882
50	10.8791	1420.5763	102.2533	10.5187	1588.2306	114.2099
60	10.9803	1920.8247	175.1958	10.9398	2154.3582	188.7208
70	10.8465	2474.9559	252.9820	10.3325	2755.5237	288.2580
80	10.9129	3098.9755	354.4388	10.7108	3414.0819	375.2574
90	10.7657	2675.1987	348.0295	10.3227	3033.8762	413.6293
100	10.9787	3006.8569	435.9619	10.6252	3333.2110	477.4898

*Time is measured in CPU Seconds.

compared to an ensemble of size 20 and a seventeen folds increase as opposed to an ensemble size of 10. Based on these observations, we conclude that increasing the number of base learners often do not benefit the learning process, while it has a detrimental effect on the allocation of computational resources.

3 Adaptive Ensemble Size Algorithm

This section details our AES algorithm, as depicted in Algorithm 1. Our method adapts the size of the ensemble, depending on the memory cost of the current window. By varying the ensemble size, our aim is to maintain the accuracy of the learning algorithm while guaranteeing high ROI [21].

The AES algorithm begins by drawing a training window, T, from a stream of evolving data, N. A total of M models (h_m) are induced independently from T. That is, we utilize an ensemble of models, h_m, from a number of random sample, $S_m \in T$, where m range from 1 to M. Recall that we use Hoeffding Trees as base learners. For a new classification instance, the predicted class with the highest number of votes is assigned to the new instance. The process samples bootstrap replicates from the data in the stream, using the Poisson probability distribution, and the models are continuously updated for each T.

At each classification instance, two variables, A and B, are maintained in order to records the average size (in bytes) of the ensemble in memory. These two parameters (A and B) are the previous and the current memory sizes of the ensemble respectively, with A initially set to zero. After each update of T, a total of M models (h_m) are induced, and the average memory size of h_m is stored in B. For every iteration of the algorithm, the values of A and B are compared. The size of the ensemble, M, is increased by 1 if $B > A$, since this implies that more memory is being utilized in the current window, compared to the previous

---

Algorithm 1. AES Algorithm

Input: N is an evolving data stream ($N \to \infty$);
$\quad\quad T$ is the training set drawn from N with examples x;
$\quad\quad Y$ is the finite set of target class values y;
$\quad\quad HT$ is the base learning algorithm (Hoeffding trees);
$\quad\quad M$ is the number of models in the ensemble, with examples drawn from T;
$\quad\quad min$ and max ensemble size values (set to 10 and 25, respectively);
$$\delta \text{ is the generalized Kronecker function: } \delta(a,b) := \begin{cases} 1 \text{ if } a = b \\ 0 \text{ if } a \neq b \end{cases}$$

1: initialize $A = 0$
2: **for all** T drawn from N **do**
3:    **for all** $m = 1$ to M **do**
4:      S_m = random sample of size d drawn from T, with replacement
5:      h_m = model induced by HT from S_m
6:      Compute average size of ensemble (bytes) $\to B$
7:      Set $k = Poisson(1)$
8:      **for all** $n = 1, 2, ..., k$ **do**
9:         Update h_m with current examples $S_m \in T$
10:    **if** $(B > A)$ & $(M < max)$ **then**
11:      $M \leftarrow M + 1$
12:    **if** $(B < A)$ & $(M > min)$ **then**
13:      $M \leftarrow M - 1$
14:    $A \leftarrow B$
Output: hypothesis: $h_{fin}(x) = argmax_{y \in Y} \sum_{m=1}^{M} \delta(y, h_m(x))$

---

window. On the other hand, the size of M is decreased by 1 if the value of B is smaller than the value of A, since it indicates that the memory utilization is lower in the current window. This process continues as the data stream evolves.

We also implemented a variant of the AES algorithm called AES-ADWIN. The main differences between AES and AES-ADWIN is that the rate of resampling of AES-ADWIN is increased and that the windows A and B are of variable-length. That is, the AES algorithm is modified so that it utilizes the ADWIN change detection algorithm [3]. Subsequently, if ADWIN detects change in the error rate of one of the models h_i then we replace the model with the highest error with a new model.

In summary, our AES and AES-ADWIN methods are based on the observation that the increase in average memory usage points towards a potential change in the data distribution, which may occur due to some form of concept drift. When this happens, we increase the ensemble size in a bid to be pro-active and to obtain a better classification result. The converse is also true. That is, a decrease in memory usage of the ensemble indicates less difficulty in model construction, so the ensemble size may potentially be reduced. The size of M is kept between the allocated maximum and minimum values, i.e. $min \leq M \leq max$. By varying the size of M as we have done in the AES algorithm, we are able to conserve resources, without having to compromise on the accuracy of the classification results, as will be shown in the next section.

4   Experimentation

We used six real life datasets from the UCI[1] Machine Learning Repository in our experimentation, namely KDD, IMDb, Forest Cover Type, Poker Hand, Electricity and Airline. All of these datasets are potentially subject to concept drift.

For example, concept drift in the Poker Hand dataset involves changing the card at hand, i.e. the poker hand. As described by [2], a training examples constitute the playing cards drawn from a standard deck of 52. Each card is described using two attributes (suit and rank), for a total of 10 predictive attributes. Note that the order of cards is important, which is why there are 480 possible Royal Flush hands instead of 4.

The IMDb dataset was derived from the Internet Movie Database (IMDb) dataset contained in the Multi-label Extension to WEKA (MEKA) repository [23]. In the data stream adaptation, it consists of binary labels 1 and 0, representing a users interest (or not) in a movie at a particular time. To simulate concept drift, the interest changes after some time. The training examples are Term Frequency-Inverse Document Frequency (TF-IDF) representation of movie annotations, to indicate how important they are to the text corpus.

The Forest Cover Type dataset [2] concerns the forest cover type for $30 \times 30\,\mathrm{m}$ cells obtained from US Forest Service (USFS) Region 2 Resource Information System (RIS). In this domain, concept drift may appear as a result of the changes in geographical conditions because of time and weather changes [19].

The airline dataset contains a record of flight schedules for commercial flights within the USA from October 1987 to April 2008. The version used in our experimentation, as well as numerous other data stream mining research, is a subset of the original dataset of size 5.76 GB containing 116 million records. The task is to predict if a flight is delayed or not and concept drift in the dataset is as a result of the changes in the flight schedules [13].

The concept drift in the Electricity dataset is shown by the changing consumption habits, event and seasons. Note that [22] showed that this dataset has the interesting property in that autocorrelation peaks at every 48 instances (24 h) due to the cycles of electricity consumption. That is, the labels are not independent; there are long consecutive periods of up and long consecutive periods of down [22].

As observed by [6], the reader should notice that, for these real datasets, we cannot unequivocally state when drifts occur. These real datasets thus serve to evaluate our algorithms in a real-life scenario rather than a concrete drift situation (Table 2).

All experiments were performed using MOA [4], with the prequential evaluation (or the so-called 'test then train') method. In this approach, the accuracy

---

[1] http://archive.ics.uci.edu/ml/.

Table 2. Summary of datasets used

Name	# of Records	# of Attributes	Characteristics
KDD'99 (10%)	494,021	42	Numeric, Nominal
Poker Hand	829,201	11	Numeric, Nominal
IMDb	120,919	1002	Numeric
Forest Covertype	581,012	55	Numeric, Nominal
Electricity	45,312	9	Numeric, Nominal
Airline	539,383	8	Numeric, Nominal

is incrementally updated and it is ensured that each example in the evaluation window is used in the learning process. It follows that this approach also does not bias the overall results to a particular window, providing a smooth accuracy increment over time. The Kappa Statistics (k^+) was used as an accuracy measure since it has been shown to be particularly well adapted to streams with concept drift [4]. Recall that we are interested in measuring the Return on Investment (ROI) of the learning process. Recently, [21] introduced a ROI measure, where the emphasis is on comparing adaptation strategies, rather than contrasting ensemble approaches. We adapted this measure to derive the mean ROI of the evaluation steps. This allows us to make direct comparisons between adapting and non-adapting algorithms and to measure the ROI in between steps of processed instances:

$$\overline{ROI} = \frac{1}{\mathcal{T} \times N} \sum_{i=1}^{\mathcal{T}} N \frac{\gamma_i}{\psi_i} = \frac{1}{\mathcal{T}} \sum_{i=1}^{\mathcal{T}} \frac{\gamma_i}{\psi_i} \tag{1}$$

where γ_i is the change in prediction accuracy (k^+); ψ_i is the cost (RAM-Hr); and $\mathcal{T}$ is the total number of windows. We used the default parameter settings throughout our experimentations. This allows us to consider the performance of our AES algorithm, without focusing on the individual learner (Hoeffding Trees).

4.1   Results

In this section, we present our experimental results against the six datasets introduced in the previous section. Firstly, we turn our attention to memory utilization, as shown in Figs. 5 and 6. The top parts of the subfigures show how the sizes of the AES and AES-ADWIN ensemble grow and shrink throughout the learning process. The bottom sections indicate the memory usage of the six classifiers we are employing, namely AES and AES-ADWIN (indicated in red and blue), as well as two variations of OzaBag and OzaBagADWIN. The first variation (Ozabag (10) and OzabagADWIN (10)) refer the default MOA settings, while the second variation refers to the ensembles with sizes equal to the number of base learners used by the AES and AES-ADWIN methods. Figures 5 and 6 clearly depicts that the sizes of the AES and AES-ADWIN ensembles vary

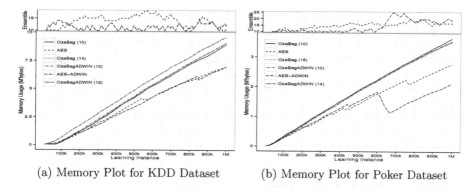

(a) Memory Plot for KDD Dataset   (b) Memory Plot for Poker Dataset

Fig. 5. Change in ensemble size and memory utilization (A)

Table 3. Kappa plus statistics median for the datasets

	KDD'99			Poker			IMDb			Forest			Electricity			Airline		
	e.s	k^+	r	e.s	k^+	r	e.s	k^+	r	e.s	k^+	r	e.s	k^+	r	e.s	k^+	r
OzaBag	10	88.0362	6	10	88.7212	4	10	87.8519	5	10	88.3543	6	10	88.4178	2	10	88.2179	4
AES	14	88.6391	2	15	89.0466	2	12	88.4434	3	14	88.8211	3	14	88.2528	4	12	88.2114	5
OzaBag	14	88.5790	3	15	88.9772	3	12	**88.8997**	1	14	**89.0140**	1	14	88.1120	5	12	88.2061	6
OzaBagADWIN	10	88.5542	4	10	88.7212	4	10	88.3559	4	10	88.3910	5	10	88.4178	2	10	88.3647	3
AES-ADWIN	12	88.0736	5	14	88.4064	5	14	87.7557	6	12	88.4043	4	15	88.3007	3	13	88.3848	2
OzaBagADWIN	12	**89.1392**	1	14	**89.2080**	1	14	88.4804	2	12	88.8891	2	15	**88.9806**	1	13	**88.4034**	1

⋆ e.s = ensemble size; r = ranking (lower is better)

considerably throughout the learning process, as based on the memory utilization within each window. The figures further show that the AES and AES-ADWIN algorithms compare favourably, in terms of memory utilization, to their OzaBag counterparts. This is the case even when the ensemble sizes increase to higher than the average values.

We further compare the methods in terms of accuracy and ROI values. In these results, we show the average ensemble sizes we obtained, rounded to the nearest whole number. We also present the values for OzaBag and OzaBagAD-WIN for these average ensemble sizes. (Recall that OzaBag and OzaBagADWIN contains a fixed, static number of base learners.) The default numbers of base learners (10) for OzaBag and OzaBagADWIN also appear. A comparison of the k^+ accuracy results using the algorithms on the different datasets is shown in Table 3.

As is clearly shown by the table, all four algorithms present highly similar levels of accuracy. The situation is quite different when the ROI is taken into account. Indeed, as shown by Table 4, our AES algorithm presents by far the best results in terms of ROI for all six (6) real life datasets. Our AES-ADWIN algorithm appears in second place.

Finally, Fig. 7 shows a graphical representation of the ROI at each step of the mining process. One first notices that, for each dataset, there is initially a surge of the ROI. Such a behavior can be traced back to the initialization of the evaluation process, when the accuracy is building up. Recall that we are

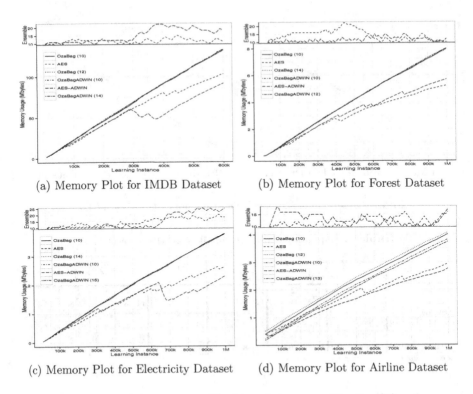

(a) Memory Plot for IMDB Dataset   (b) Memory Plot for Forest Dataset

(c) Memory Plot for Electricity Dataset   (d) Memory Plot for Airline Dataset

Fig. 6. Change in ensemble size and memory utilization (B)

Table 4. ROI comparison between the algorithms

	KDD'99			Poker			IMDb			Forest			Electricity			Airline		
	e.s	ROI	r	e.s	ROI	r	e.s	ROI	r	e.s	ROI	r	e.s	ROI	r	e.s	ROI	r
AES	14	**1.7207**	1	15	**4.0195**	1	12	**1.3558**	1	14	**0.6184**	1	14	**1.2899**	1	12	**0.3871**	1
OzaBag	14	0.6557	3	15	2.2341	3	12	0.7991	3	14	0.4436	3	14	0.6790	3	12	0.1988	4
AES-ADWIN	12	0.7526	2	14	3.4481	2	14	0.8425	2	12	0.3732	4	15	1.1231	2	13	0.2344	3
OzaBagADWIN	12	0.3103	4	14	1.9285	4	14	0.5285	4	12	0.4511	2	15	0.5813	4	13	0.2477	2

⋆ e.s = ensemble size; r = ranking (lower is better)

using the prequential evaluation method. This indicates that there is a rapid buildup in accuracy at this point, before the latter peaks. Then, the ROI rapidly decreases, and remains essentially constant.

In summary, our results indicate that varying the ensemble size during online learning is beneficial, in that the amount of resources employed does not have a detrimental effect on the error rates. That is, the AES approach yields comparable results, in terms of accuracies, while the overall resource utilization, measured in terms of model construction time and memory utilization, is lowered. We believe that this initial result is worth exploring further. We are especially interested in extending our work to the area of mobile data mining, where there

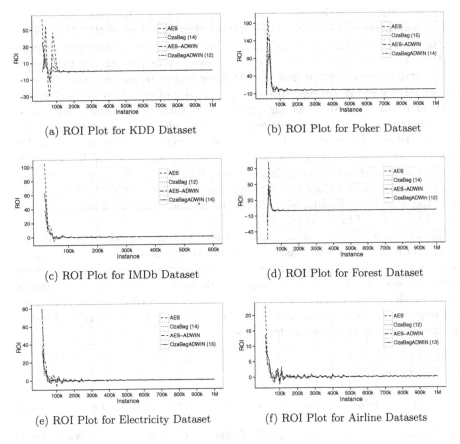

(a) ROI Plot for KDD Dataset

(b) ROI Plot for Poker Dataset

(c) ROI Plot for IMDb Dataset

(d) ROI Plot for Forest Dataset

(e) ROI Plot for Electricity Dataset

(f) ROI Plot for Airline Datasets

Fig. 7. ROI plots for electricity and airline datasets

is a need for real-time data analytics on small devices that are susceptible to varying degrees and types of concept drift [6,16].

5   Conclusion

In a data stream setting, where we are potentially dealing with massive, fast evolving data, it is important to consider the overall utility of the learning process. This is especially relevant in a scenario where the resources are limited, for instance mobile data mining applications. This paper presented an online ensemble-based approach that adapts the size of an Online Bagging ensemble during learning from data streams. This property allows for flexibility in terms of resource usage and yields a high Return on Investment.

While our initial results are promising, a number of research avenues remain. Our next step will be to extend our work to specific case studies within the mobile data mining scenario. Within this setting, there is a need for light-weight algorithms that not only limit computational resources, but also consider the screen real-estate and energy considerations of mobile devices [16]. We plan to extend our AES algorithm to incorporate hybrid adaptation strategies, that are both resource-aware and situation-aware, in order to address this challenge [12].

We will further explore how the AES algorithm would perform when the types and degrees of concept drifts vary. The use of our AES method in seasonal or re-occurring drift settings, in domains such as economics, retail sales or electricity consumption, needs our future consideration. Here, we are interested in determining whether the use of prior memory patterns may be used to predict near-future resources needs. Further experimentation will be done with more artificial datasets, with different kinds of concept drifts (sudden, gradual, seasonal and mixed) induced at varying time intervals. Exploring other intelligent update rules and using parameter settings other than the default values are another avenues of future research.

In our current work, we did not consider the impact of noise. It will be important to take this into consideration. The reactions of the utilization cost to noise, and how to harness this to improve accuracy is another area of research worth considering. We intend to explore this by extending our algorithms to minimise the effects of noise in the stream. We will also consider the noise ratio before processing any stream window, so that we can get the best result possible in that processing cycle, at the least cost.

References

1. Attar, V., Sinha, P., Wankhade, K.: A fast and light classifier for data streams. Evol. Syst. **1**(3), 199–207 (2010)
2. Bifet, A., Holmes, G., Pfahringer, B., Kirkby, R., Gavaldà, R.: New ensemble methods for evolving data streams. In: Proceedings of the 15th ACM SIGKDD International Conference on Knowledge Discovery and Data Mining, KDD 2009, NY, USA, pp. 139–148 (2009)
3. Bifet, A., Holmes, G., and Pfahringer, B.: Leveraging bagging for evolving data streams. In: Proceedings of the European Conference on Machine Learning and Principles and Practice of Knowledge Discovery in Databases, ECML/PKDD, pp. 135–150 (2010)
4. Bifet, A., Holmes, G., Kirkby, R., Pfahringer, B.: MOA: massive online analysis. J. Mach. Learn. Res. **11**, 1601–1604 (2010)
5. Breiman, L.: Bagging predictors. Mach. Learn. **24**(2), 23–140 (1996)
6. Brzezinski, D., Stefanowski, J.: Reacting to different types of concept drift: the accuracy updated ensemble algorithm. IEEE Trans. Neural Netw. Learn. Syst. **25**(1), 81–94 (2014)
7. Datar, M., Gionis, A., Indyk, P., Motwani, R.: Maintaining stream statistics over sliding windows. In: 13th Annual ACM-SIAM Symposium on Discrete Algorithms, pp. 635–644 (2002)
8. Domingos, P., Hulten, G.: Mining high-speed data streams. In: Proceedings of the Sixth ACM SIGKDD International Conference on Knowledge Discovery and Data Mining, KDD 2000, NY, USA, pp. 71–80 (2000)

9. Gama, J., Rocha, R., Medas, P.: Accurate decision trees for mining high-speed data streams. In: Proceedings of the Ninth ACM SIGKDD International Conference on Knowledge Discovery and Data Mining, KDD 2003, pp. 523–528 (2003)

10. Gaber, M.M., Stahl, F., Gomes, J.B.: Pocket Data mining: Big Data on Small Devices. Studies in Big Data, vol. 2. Springer, Heidelberg (2014)

11. Hansen, L.K., Salamon, P.: Neural network ensembles. IEEE Trans. Pattern Anal. Mach. Intell. 12(10), 993–1001 (1990)

12. Haghighi, P.D., Zaslavsky, A., Krishnaswamy, S., Gaber, M.M., Loke, S.: Context-aware adaptive data stream mining. Intell. Data Anal. 13(3), 423–434 (2009)

13. Ikonomovska, E.: Airline dataset (2011). http://kt.ijs.si/elena_ikonomovska/data.html. Accessed 20 Jan 2015

14. Kargupta, H., Hoon, P., Pittie, S., Liu, L.: Mobimine: monitoring the stock market from a PDA. ACM SIGKDD Explor. 3, 37–47 (2002)

15. Kolter, J.Z., Maloof, M.A.: Dynamic weighted majority: an ensemble method for drifting concepts. J. Mach. Learn. Res. 8, 2755–2790 (2007)

16. Krishnaswamy, S., Gama, J., Gaber, M.M.: Mobile data mining: from algorithms to applications. In: IEEE 13th International Conference on Mobile Data Management (MDM), pp. 360–363 (2012)

17. Opitz, D., Maclin, R.: Popular ensemble methods: an empirical study. J. Artif. Intell. Res. 11, 169–198 (1999)

18. Oza, N.C., Russell, S.: Online bagging and boosting. In: Artificial Intelligence and Statistics, pp. 105–112 (2001)

19. Oza, N.C., Russell, S.: Experimental comparisons of online and batch versions of bagging and boosting. In: Proceedings of the Seventh ACM SIGKDD International Conference on Knowledge Discovery and Data Mining, KDD 2001, pp. 359–364 (2001)

20. van Rijn, J.N., Holmes, G., Pfahringer, B., Vanschoren, J.: Algorithm selection on data streams. In: Džeroski, S., Panov, P., Kocev, D., Todorovski, L. (eds.) DS 2014. LNCS, vol. 8777, pp. 325–336. Springer, Heidelberg (2014)

21. Žliobaite, I., Budka, M., Stahl, F.: Towards cost-sensitive adaptation: when is it worth updating your predictive model? Neurocomputing 150, 240–249 (2015)

22. Žliobaite, I.: How good is the Electricity benchmark for evaluating concept drift adaptation. arXiv preprint arXiv: 1301.3524 (2013)

23. Žliobaite, I., Bifet, A., Pfahringer, B., Holmes, G.: Active learning with drifting streaming data. IEEE Trans. Neural Netw. Learn. Syst. 25(1), 27–39 (2014)

Mining Periodic Changes in Complex Dynamic Data Through Relational Pattern Discovery

Corrado Loglisci[1,2(✉)] and Donato Malerba[1,2]

[1] Department of Computer Science, Universita' degli Studi di Bari "Aldo Moro",
Bari, Italy
{corrado.loglisci,donato.malerba}@uniba.it
[2] CINI- Consorzio Interuniversitario Nazionale per l'Informatica, Bari, Italy

Abstract. The empowerment of the information technologies in many real-world applications has opened to the possibility of tracking complex and evolving phenomena and gather information able to describe such phenomena. For instance, in bio-medical applications, we can monitor a patient and collect data that range from his clinical picture to the laboratory studies on biological products. In this scenario, studying the possible alterations manifested over time becomes thus relevant and, in life sciences, even determinant. In this paper, we investigate the task of determining changes which are regularly repeated over time and we propose a method based on two notions of patterns, *emerging patterns* and *periodic changes*. The method works on a time-window model to the end of *(i)* capturing statistically evident changes and *(ii)* detecting their periodicity. The method was applied to two typical real-world scenarios with complex dynamic data, that is, Virology and Meteorology.

1 Introduction

The advances in the development of hardware devices and information technologies has augmented the complexity of the applications in fields such as life sciences, engineering and social sciences. The most prominent result lies in the possibility of following and dealing with complex phenomena, whose investigation passes through the analysis of the data that being generated and collected. Such data are high heterogeneous, characterized by different properties and composed of several entities. Another degree of complexity is represented by the fact that these applications often work in evolving scenarios, which leads to generate data dynamic in nature. The analysis of complex and time-changing data can provide us with tools able to deal with unexpected behaviors and understanding the underlying dynamics. As proof of that, we can mention an extensive list of techniques of data mining focused on complex dynamic data.

Mining changes in complex data is not a problem of immediate solution and we could not directly apply traditional methods to such data because the existing approaches would tend to oversimplify the different sources of information of the complex data. Indeed, changes in complex data can be originated from multiple and different factors, such as the structure, descriptive properties, entities

© Springer International Publishing Switzerland 2016
M. Ceci et al. (Eds.): NFMCP 2015, LNAI 9607, pp. 76–90, 2016.
DOI: 10.1007/978-3-319-39315-5_6

constituting the main complex objects. This seems to be the same consideration which has stimulated recent works of change mining on structured evolving
data. For instance, the method proposed in [13] analyzes dynamic networks in
order to detect changes occurred at the level the labels of the edges, while in
[12], the same authors consider again dynamic networks but with the different
task of identifying changes in the number of the occurrences of sub-networks.
Changes can be searched also as variations of the global and local properties [4]
of evolving graphs.

In many applications, changes can be unpredictable, they can reveal unexpected variations or can be even evolutions already observed in past which are
newly repeated. The analysis of repetitive behaviors exhibited over time is not
a novel problem and the literature on the periodicity detection contributes to
it. The blueprint for most algorithms follows a frequent pattern-based framework according which repetitive behaviors can be identified as regularly (periodically) repeated sub-sequences in a lengthy sequence [6]. A common aspect
to these works is that the periodic repetitions concerns stationary data (e.g.,
sub-sequences) that are replicated over time, while no attempt has been done
for detecting periodic changes, that is, dynamic behaviors which regularly recur.

The identification of periodic changes cannot be trivially faced by resorting
to any algorithm of discovery of periodic patterns due to several reasons. First,
the periodic patterns describe repeated behaviors which are static, thus their
periods would not refer to changes. Second, searching changes among the periodic
patterns could lead to work on a restricted search space. Third, different periodic
patterns could depict changes but their periods could be not aligned, therefore
the problem would require further computation to determine the periodicity of
the changes from the periodicity of the patterns. Fourth, the existing algorithms
work on a sequence-based representation, which, in the case of complex data,
tends to over-simplify the multiple and different aspects of the data with the risk
of omitting periodic behaviors and changes originating from the inner entities.

In this paper, we investigate the task of capturing statistically evident
changes emerged over time and tracking their repeatability. The proposed
method adopts a model of analysis based on time-windows and uses a frequent
pattern mining framework as mean for abstracting and summarizing the data.
This enables us to search changes as differences between frequent patterns. Since
frequency denotes regularity, patterns can provide empirical evidence about real
changes. Frequent patterns are discovered from the complex data collected by
time-windows and thus they reflect co-occurrences in terms of structure, properties and inner entities of the complex objects that are frequent in specific
intervals of time. The changes which can emerge in this setting regard differences between the frequent patterns of two time-windows. In particular, we are
interested in changes which are manifested as significant variations of the frequency of the patterns from a time-window to the next one. Not all the changes
are considered, but only those which are replicated over time. We extend the concept of *Emerging Patterns* in order to depict changes between two time-windows

and introduce the notion of *Periodic Changes* in order to characterize changes regularly repeated over time.

The paper is organized as follows. In Sect. 2 we report necessary notions, while the method is described in Sect. 3. An application to the phenotype data in Virology is described in Sect. 4. Then, we overview the related literature (Sect. 5) and finally conclusions close the paper (Sect. 6).

2   Basics and Definitions

Most of the methods reported in the literature overcome the difficulty of accounting for the different aspects of the complex data objects with formalisms based on vectors or attribute-value sets, which model only global properties of the data. These solutions could be too limiting because they tend to neglect both the intrinsic structure of the complex data and the entities that constitute the main complex object and the inner relationships existing among the entities. To overcome this drawback, we use the (multi-) relational setting, which has been argued to be the most suitable formalism for representing complex data, since it can deal with the heterogeneity and it can naturally represent a large variety of relationships among data.

In the relational setting, complex data and the constituting entities can play different roles in the analysis. We can distinguish them between target objects (TOs) and non-target objects ($NTOs$). The former are the main subjects of the analysis, while the latter are objects relevant for the current problem and associated with the former.

Let $\{t_1 \ldots t_n\}$ be a a sequence of time-points. At each time-point t_i, a set of instances (TOs) is collected. A *time-window* τ is a sequence of consecutive time-points $\{t_i, \ldots, t_j\}$ ($t_1 \leq t_i$, $t_j \leq t_n$) which we denote as $[t_i; t_j]$. The width w of a time-window is the number of time-points in τ, i.e. $w = j - i + 1$. Two time-windows τ and τ' defined as $\tau = [t_i, ; t_{i+w-1}]$ and $\tau' = [t_{i+w}; t_{i+2w-1}]$ are *consecutive*.

Let $\tau = [t_i; t_{i+w-1}]$, $\tau' = [t_{i+w}; t_{i+2w-1}]$, $\tau'' = [t_j; t_{j+w-1}]$, and $\tau''' = [t_{j+w}; t_{j+2w-1}]$) be time-windows, two pairs of consecutive time-windows (τ, τ') and (τ'', τ''') are δ-*separated* if $(j + w) - (i + w) \leq \delta$ ($\delta > 0$, $\delta \geq w$). Two pairs of consecutive time-windows (τ, τ') and (τ'', τ''') are *chronologically ordered* if $(j + w) > (i + w)$. We assume that all the time-windows have the same width w. In the remaining of the paper, we use the notation τ_{h_k} to refer to a time-window and the notation (τ_{h_1}, τ_{h_2}) to indicate a pair of consecutive time-windows.

Both TOs and $NTOs$ can be represented in the Datalog language as sets of ground atoms. A ground atom is an n-ary logic predicate symbol applied to n constants. We consider three categories of logic predicates: (1)*key predicate*, which identifies the TOs, (2)*property predicates*, which define the value taken by a property of a TO or of a $NTOs$, and (3)*structural predicates*, which relate TOs with their $NTOs$ or relate the $NTOs$ each other.

In the following, we report an example of TO in the domain of virology. Virology is one of the fields in which rich sets of complex dynamic data can be collected.

virus(p1). epidemiological_condition(p1,enhanced_Trasmission_to_Human).
epidemiological_condition(p1,severity). epidemiological_condition(p1,increased_Virulence).
clinical_condition(p1,adamantane_resistance). clinical_condition(p1,oseltamivir_resistance).
clinical_condition(p1,polybasic_HA_Cleavage).dependent_by(adamantane_resistance,mutation).
variation_of(mutation,protein_M2). is_a(mutation,v26F).
where

- *virus()* is the key predicate which identifies the TO named as $p1$;
- *epidemiological_condition(_,_)*, *clinical_condition(_,_)* are structural predicates which relates the TO with the $NTOs$ identified as *enhanced_Trasmission_to_Human, severity, increased_Virulence, adamantane_resistance, oseltamivir_resistance,polybasic_HA_Cleavage*;
- *dependent_by(_,_)* is a structural predicate which relates the NTO *adamantane_resistance* with the NTO identified as *mutation*;
- *variation_of(_,_)* is a structural predicate which relates the NTO *mutation* to the NTO identified as *protein_M2*
- *is_a(_,_)* is a property predicate which relates the NTO *mutation* to the value v26F

The following definitions are crucial for this work:

Definition 1. Relational pattern
    A conjunction of atoms $P = p_0(t_0^1), p_1(t_1^1, t_1^2), p_2(t_2^1, t_2^2), \ldots, p_m(t_m^1, t_m^2)$, is a relational pattern if p_0 is the key predicate, p_i, $i = 1, \ldots, m$ is either a structural predicate or a property predicate. ∎

Terms t_i^j are either constants, which correspond to values of the property predicates, or variables, which identify TOs or $NTOs$. Moreover, all variables are linked to the variable used in the key predicate (according to the linkedness property [14]).
    A relational pattern P is characterized by a statistical parameter, namely the *support* (denoted as $sup_{\tau_{h_k}}(P)$), which denotes the relative frequency of P in the time-window τ_{h_k}. It is computed as the number of TOs of the time-window τ_{h_k} in which P occurs divided by the total number of TOs of τ_{h_k}. When the support exceeds a minimum user-defined threshold $minSUP$, P is said to be *frequent* in the time-window τ_{h_k}.

Definition 2. Emerging pattern-EP
    Let (τ_{h_1}, τ_{h_2}) be a pair of consecutive time-windows; P be a frequent relational pattern in the time-windows τ_{h_1} and τ_{h_2}; $sup_{\tau_{h_1}}(P)$ and $sup_{\tau_{h_2}}(P)$ be the support of the pattern P in τ_{h_1} and τ_{h_2} respectively. P is an emerging pattern in (τ_{h_1}, τ_{h_2}) iff $\frac{sup_{\tau_{h_1}}(P)}{sup_{\tau_{h_2}}(P)} \geq minGR$ $\vee$ $\frac{sup_{\tau_{h_2}}(P)}{sup_{\tau_{h_1}}(P)} \geq minGR$ ∎

where, $minGR$ (>1) is a user-defined minimum threshold.
    The ratio $sup_{\tau_{h_1}}(P)/sup_{\tau_{h_2}}(P)$ ($sup_{\tau_{h_2}}(P)/sup_{\tau_{h_1}}(P)$) is denoted with $GR_{\tau_{h_1},\tau_{h_2}}(P)$ ($GR_{\tau_{h_2},\tau_{h_1}}(P)$) and it is called *growth-rate* of P from τ_{h_1} to τ_{h_2} (from τ_{h_2} to τ_{h_1}). When $GR_{\tau_{h_1},\tau_{h_2}}(P)$ exceeds $minGR$, the support of P

decreases from τ_{h_1} to τ_{h_2} by a factor equal to the ratio $sup_{\tau_{h_1}}(P)/sup_{\tau_{h_2}}(P)$, while when $GR_{\tau_{h_2},\tau_{h_1}}(P)$ exceeds $minGR$, the support of P increases by a factor equal to $sup_{\tau_{h_2}}(P)/sup_{\tau_{h_1}}(P)$.

The concept of emerging pattern is not novel in the literature [3]. In its classical formulation, it refers to the values of support of the same pattern which has been discovered in two different classes of data, while, in this work, we extend it to represent the differences between the data collected in two intervals of time, and therefore, we refer to the values of support of the same pattern which has been discovered in two time-windows.

In the following, we report an example of emerging pattern. the time-points correspond to the years. Consider the relational pattern P

P: *virus(H),clinical_condition(H,C),dependent_by(C,M),variation_of(M,N)*.

with τ_{h_1}=[1991;1995], τ_{h_2}=[1996;2000], $sup_{[1991;1995]}(P) = 0.8$ and $sup_{[1996;2000]}(P) = 0.5$. Here, the support of the pattern P decreases, whereby of the growth-rate $GR_{[1991;1995],[1996;2000]}(P)$ is 1.6 (0.8/0.5). By supposing that $minGR$=1.5, the pattern P is considered emerging in ([1991;1995],[1996;2000]).

Definition 3. Periodic change-PC

Let $T : \langle(\tau_{i_1},\tau_{i_2}),\ldots,(\tau_{m_1},\tau_{m_2})\rangle$ be a set of chronologically ordered pairs of time-windows; P be an emerging pattern between the time-windows τ_{h_1} and τ_{h_2} for all $h = i,\ldots,m$; $\langle GR_{\tau_{i_1},\tau_{i_2}},\ldots,GR_{\tau_{m_1},\tau_{m_2}}\rangle$ be the values of growth-rate of P in the pairs $\langle(\tau_{i_1},\tau_{i_2}),\ldots,(\tau_{m_1},\tau_{m_2})\rangle$ respectively; $\Theta_P : \Re \to \Psi$ be a function which maps $GR_{\tau_{h_1},\tau_{h_2}}(P)$ into a discrete value $\psi_{\tau_{h_1},\tau_{h_2}} \in \Psi$ for all $h = i,\ldots,m$. P is a periodic change iff:

1. *$|T| \geq minREP$*
2. *(τ_{h_1},τ_{h_2}) and (τ_{k_1},τ_{k_2}) are δ-separated for all $h = i,\ldots,m-1$, k=h+1 and there is no pair (τ_{l_1},τ_{l_2}), $h < l$, s.t. (τ_{h_1},τ_{h_2}) and (τ_{l_1},τ_{l_2}) are δ-separated*
3. *$\psi = \psi_{\tau_{i_1},\tau_{i_2}} = \ldots = \psi_{\tau_{m_1},\tau_{m_2}}$* ∎

where $minREP$ is a minimum user-defined threshold. A PC is a frequent pattern whose support increases (decreases) at least $minREP$ times with an order of magnitude greater than $minGR$. Each change (increase/decrease) occurs within δ time-points and it is characterized by a discrete value in the set Ψ. Intuitively, a PC represents a variation, manifested with a particular periodicity, of the support of the same pattern.

Note that Definition 3 uses the threshold δ as maximum value of periodicity, which leads to a two-fold result. On one hand, the periodicity of the PCs may have any value lower than δ and therefore the detected periodic changes could refer to EPs which are not repeated with a perfect regularity. On the other hand, this allows us to discover a larger set of periodic changes that includes also those that are not perfectly recurring, that is, changes with some disturbances between the repetitions. In the literature, this kind of periodicity is referred as *asynchronous*[7].

An example of PC is reported here. Consider the following EPs

P: *virus(H),clinical_condition(H,C),dependent_by(C,M), variation_of(M,N)*
emerging in ([1991;1992],[1993,1994])

P: *virus(H),clinical_condition(H,C),dependent_by(C,M), variation_of(M,N)*
emerging in ([1996;1997],[1998;1999])

P: *virus(H),clinical_condition(H,C),dependent_by(C,M), variation_of(M,N)*
emerging in ([1999;2000],[2001;2002])

P: *virus(H),clinical_condition(H,C),dependent_by(C,M), variation_of(M,N)*
emerging in ([2004;2005],[2006,2007])

Here, $\psi_{[1991;1992],[1993;1994]} = \psi_{[1996;1997],[1998;1999]} = \psi_{[2004;2005],[2006:2007]}$, $\psi_{[1991;1992],[1993;1994]} \neq \psi_{[1999;2000],[2001;2002]}$. By supposing $minREP{=}2$ and $\delta = 6$, P is a periodic change. Indeed, $T : \langle([1991;1992],[1993;1994]),([1996;1997], [1998,1999])\rangle$ meets the conditions (1) and (2) because $|T| = 2$ and (1998-1993)<6. The discrete values of the growth-rate in ([1991;1992],[1993;1994]) and ([1996;1997],[1998;1999]) meet the condition (3). The pair of time-windows ([1999;2000],[2001;2002]) is not considered because $\psi_{[1999;2000],[2001;2002]}$ does meet the condition (3), while the pair of time-windows ([2004;2005],[2006;2007]) does not meet the condition (2) because (2006-1998)>6.

3 The Method

We propose a method which discovers PCs incrementally as time goes by (Fig. 1). It works on the succession $\langle(\tau_{1_1},\tau_{1_2}),\ldots,(\tau_{h_1},\tau_{h_2}),\ldots,(\tau_{z_1},\tau_{z_2})\rangle$ of pairs of time-windows obtained from $\{t_1,\ldots,t_n\}$. Each time-window τ_{u_v} (except the first and last one) is present in two consecutive pairs, more precisely, the pair (τ_{h_1},τ_{h_2}) where $\tau_{u_v} = \tau_{h_2}$, and the pair $(\tau_{(h+1)_1},\tau_{(h+1)_2})$ with $\tau_{u_v} = \tau_{(h+1)_1}$. This is done with the intent to capture the changes of support of the patterns from τ_{h_1} to τ_{u_v} and from τ_{u_v} to $\tau_{(h+1)_2}$. The method performs three steps:

(i) Discovery of frequent patterns (FPs) for each time-window τ_{u_v};

(ii) Extraction of the EPs by matching the frequent patterns discovered from τ_{h_1} against the frequent patterns discovered from τ_{h_2}. These EPs are stored in a pattern base, which is incrementally updated as the time-windows are processed;

(iii) Detection of PCs by testing the conditions of Definition 3 on the EPs stored in the base.

Note that, when the algorithm processes the pair $(\tau_{(h+1)_1},\tau_{(h+1)_2})$, it uses the frequent patterns of the time-window τ_{h_2}, which had been discovered when the algorithm had processed the pair (τ_{h_1},τ_{h_2}). This avoids of performing the step (i) twice on the same time-window. Details on these three steps are reported in the following.

3.1 Relational Frequent Pattern Discovery

Frequent patterns are discovered from each time-window by following the method proposed in [13], which enables the discovery of relational patterns whose support exceeds $minSUP$. It explores level-by-level the lattice of the patterns, from

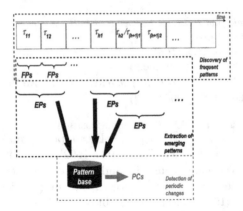

Fig. 1. The block-diagram of the three-step method for detecting periodic changes.

the most general to the more specific ones, starting from the most general pattern (which contains only the key predicate). The lattice is organized according to a generality ordering based on the notion of θ-subsumption [14]. Formally, given two relational patterns $P1$ and $P2$, we say that "$P1$ ($P2$) is more general (specific) than $P2$ ($P1$) under θ-subsumption" if and only if "$P2$ θ-subsumes $P1$", where $P2$ θ-subsumes $P1$ if and only if a substitution θ exists such that $P2$ $\theta \subseteq P1$. We denote this relation of generality of $P1$ with respect to $P2$ as $P1 \geqslant_\theta P2$.

The method implements a two-stepped procedure: *(i)* generation of candidate patterns with k atoms (k-th level) by considering the frequent patterns with $k-1$ atoms ((k-1)-th level); *(ii)* evaluation of the support of the patterns with k atoms.

The monotonicity property of the support value (i.e., a super-set of an non-frequent pattern cannot be frequent) is exploited to avoid the generation of non-frequent relational patterns. In fact, in accordance with the Definition 2, non-frequent patterns are not used for detecting changes and thus we can prune portions of the space containing non-frequent patterns. Thus, given two relational patterns $P1$ and $P2$ with $P1 \geqslant_\theta P2$, if $P1$ is non-frequent in a time-window, then the support of $P2$ is less than the threshold $minSUP$ and it is non-frequent too in the same time-window. Therefore, we do not refine the patterns which are non-frequent.

3.2 Emerging Pattern Extraction

Once the frequent patterns have been discovered from the time-windows τ_{h_1} and τ_{h_2}, they are evaluated in order to check if the growth-rate exceeds the threshold $minGR$. Unfortunately, the monotonicity property does not hold for the growth-rate. In fact, given two frequent patterns $P1$ and $P2$ with $P1 \geqslant_\theta P2$, if $P1$ is not emerging, namely $GR_{\tau_{h_1}, \tau_{h_2}}(P1) < minGR$ ($GR_{\tau_{h_2}, \tau_{h_1}}(P1) < minGR$), then the pattern $P2$ may or may not be an EP, namely its growth-rate could exceed the threshold $minGR$. However, we can equally optimize this step by acting on the evaluation of some patterns. Indeed, we avoid the evaluation of the refinements of a pattern P discovered from the time-window τ_{h_1} (τ_{h_2}) whenever P is non-frequent in the time-window τ_{h_2} (τ_{h_1}). Note that this operation could exclude EPs with very high values of growth-rate (i.e., the strongest changes), but here we are interested in the changes exhibited by co-occurrences which are statistically evident in both intervals of time.

The EPs extracted on the pairs of time-windows are stored in the pattern base, which hence contains the frequent patterns that satisfy the constraint set by $minGR$ on at least one pair of time-windows. Each EP is associated with two lists, named as $TWlist$ and $GRlist$. $TWlist$ is used to store the pairs of time-windows in which the growth-rate of the pattern exceeds $minGR$, while $GRlist$ is used to store the corresponding values of growth-rate. To distinguish the changes due to the decrease of the support from those due to the increase, we store the values of growth-rate as negative (that is, with the minus sign) when it decreases.

The pattern base is maintained with two operations, namely insertion of the EPs and update of the lists $TWlist$ and $GRlist$ associated with the EPs. A pattern is inserted if it has not been recognized as emerging in the previous pairs of time-windows, while, if it has been previously inserted, we update the two lists.

3.3   Periodic Change Detection

The step (3) works on the pattern base and filters out the EPs that do not meet the conditions of Definition 3. The function Θ_P implements an equal-width discretization technique, which is applied to two sets of values obtained from the lists $GRlist$ of all the stored EPs, the first set consists of all the positive values of growth-rate, the second set consists of all the negative values. Note that we have not infinite values of growth-rate because all the patterns considered are frequent, i.e., there are no values of support equal to zero. The ranges returned by the discretization technique correspond to the discrete values $\psi_{\tau_h,\tau_{h+1}}$. Thus, we have two sets of ranges Ψ^+ and Ψ^-: Ψ^+ refers to the discrete values obtained from the positive values of growth-rate, while Ψ^- refers to the discrete values obtained from the negative values. We replace the numeric values contained in the lists $GRlist$ with the corresponding ranges in Ψ^+ and Ψ^-. This allows us to obtain two separate sets of discrete values and capture the increases/decreases of the support of the patterns by representing them with a finite number of cases.

The discretization technique tends to collect growth-rate values that are closer in the same range (discrete value). This could suggest to reduce the number of EPs that have the same discrete value by pruning the EPs that are more general (i.e., those with a lower number of predicates) and conserving the EPs that are more specific (i.e., those with a higher number of predicates). But, this cannot be done because it is not guaranteed that patterns which have the same discrete values in some time-windows continue to maintain the same values in the others time-windows.

Data: EPs: the base of the emerging patterns discovered from the pairs of
time-windows $\langle (\tau_{1_1}, \tau_{1_2}), \ldots, (\tau_{h_1}, \tau_{h_2}), \ldots, (\tau_{z_1}, \tau_{z_2}) \rangle$
Data: $minREP$: the minimum threshold of the number of repetitions
Data: δ: the maximum threshold of periodicity
Result: PCs: the set of valid periodic changes

1 **for** $\rho \in \langle (\tau_{1_1}, \tau_{1_2}), \ldots, (\tau_{h_1}, \tau_{h_2}), \ldots, (\tau_{z_1}, \tau_{z_2}) \rangle$ **do**

2 $\rho\_EPs = getEmerging(EPs, \rho)$ // Return the emerging patterns
extracted from the pair of time-windows ρ

3 **for** $EP \in \rho\_EPs$ **do**

4   **for** $\overline{\psi} \in (\Psi^+ \cup \Psi^-)$ **do**

5   **if** $(contained(EP.TWlist, \rho)$ and $equal(getDiscrete(EP.GRlist), \rho), \overline{\psi})$
// Check whether EP is emerging in ρ and it has $\overline{\psi}$
as discrete value

6   **then**

7              $candidatePC \leftarrow getCandidate(candidates, EP, \overline{\psi})$ // Return the
candidate PC associated with EP that has $\overline{\psi}$ as
discrete value

8              $last \leftarrow getLastPair(candidatePC.T)$

9   **if** $(\delta - separated(last, \rho))$ **then**

10                 $append(candidatePC.T, \rho)$

11                 $update(candidates, candidatePC)$

12   **else**

13                 $insert(completePCs, candidatePC)$

14                 $remove(candidates, candidatePC)$

15                 $newCandidate.\psi = \overline{\psi}$

16                 $newCandidate.T = newCandidate.T \cup \rho$ // Build the set
T of a new candidate PC with ρ

17                 $insert(candidates, newCandidate)$

18 $completePCs = completePCs \cup candidates$

19 $PCs = filterByMinREP(completePCs)$

Algorithm 1: Detection of Periodic Changes

In this step, the method performs two preliminary operations: (i) removal of
the EPs whose lists $TWlist$ and $GRlist$ have less than $minREP$ elements; (ii)
sorting the remaining lists $TWlist$ by chronological order. The lists $GRlist$ will
be re-arranged accordingly.

The method discovers PCs by working on the EPs separately and it can detect
more than one PC from a single EP. The algorithmic description is reported in
Algorithm 1. As the succession of pairs of time-windows is scanned, the algorithm
considers the list $TWlist$ for each EP and evaluates the pairs of time-windows
where EP has the same discrete value $\overline{\psi}$ (line 5). The used pairs of time-windows
and the discrete value $\overline{\psi}$ define a candidate PC. Thus, the algorithm evaluates
the current pair of time-windows ρ of $TWlist$ against with the latest pair inserted
in the set T of the candidate PC that has the same discrete value (lines 6–9): if
the pairs of time-windows are δ-separated, then the pair ρ is inserted in the set
T of the candidate PC (lines 10–11), otherwise it can be considered to start the

construction of the set T of a new candidate PC having the same discrete value $\bar{\psi}$ (lines 12–17). Finally, the PCs whose sets T have less than $minREP$ pairs are filtered out (line 19).

In order to clarify how the step *(3)* works, we report an explanatory example. Consider $\Psi^+ = \{\psi', \psi''\}$, $minREP=3$, $\delta=13$ and the lists $TWlist$ and $GRlist$ built as follows:

$TWlist$: $\langle$([1970; 1972], [1973; 1975]) , ([1976; 1978], [1979; 1981]) , ([1982; 1984], [1985, 1987]) ,
         ([1988; 1990], [1991; 1993]) , ([1994; 1996], [1997; 1999]) , ([2010; 2012], [2013; 2015])$\rangle$

$GRlist$: $\langle$ ψ', ψ', ψ'',

                 ψ', ψ'', $\psi'\rangle$

By scanning the list $TWlist$, we can initialize the set T of a candidate PC' by using the pairs ([1970;1972],[1973;1975]) and ([1976;1978],[1979;1981]) since they are δ-separated (1979-1973$<\delta$) and they have the same discrete value ψ'. The pair ([1982;1984],[1985;1987]) instead refers to a different discrete value (ψ'') and therefore it cannot be inserted into T of PC'. We use it to initialize the set T of a new candidate PC", which thus will include the time-windows referred to ψ''. Subsequently, the pair ([1988;1990],[1991;1993]) is inserted into T of PC' since its distance from the latest pair is less than δ (1991-1979$<\delta$). Then, T of PC" is updated with ([1994;1996],[1997;1999]) since 1997-1985 is less than δ, while the pair ([2010;2012],[2013;2015]) cannot be inserted into T because the distance between 2013 and 1997 is greater than δ. Thus, we use the pair ([2010;2012],[2013;2015]) to initialize the set T of a new candidate PC"'. The set T of PC' cannot be further updated, but, since its size exceeds $minREP$, we consider the candidate PC' as valid periodic change. Finally, the candidate PC" cannot be considered valid since its size is less than $minREP$. The candidate PC"' is not even considered since $|T_{\psi'}| < minREP$.

4   Experiments

We prove the viability of the proposed method on two real-world scenarios in Life Sciences. The first scenario is in Virology, where we consider the data of the phenotypes of the influenza A/H1N1 virus. Phenotypes include information at the level of proteins, gene sequences, clinical status and epidemiological factors. The analysis of the phenotype data from a perspective based on time is not novel [10,11], what is instead innovative in this work is the study of the periodicity. The second scenario is in Meteorology, where our analysis focuses on the evolution of the weather on a specific geographic area. The weather include information on the atmospheric and physical parameters. We present these two case studies separately.

Virology. Studying the alterations of the phenotypes on a temporal basis is relevant and even determinant whether considering the biological re-assortment between the involved organisms and the cyclic nature of the pandemic outbreaks. The influenza A virus can infect several species. The sources of complexity of such data are several. In particular, the virus contains eight segments gene of negative single-stranded RNA, namely $PB2$, $PB1$, PA, HA, NP, NA, M, and NS encoding for 11 proteins. Moreover, the subtype of influenza A virus is determined by the antigenicity (the capacity to induce an immune response) of the two surface glycoproteins, haemagglutinin (HA) and neuraminidase (NA). Distinct subtypes determining different clinical and epidemiological conditions [5].

The datasets we use comprise phenotype data describing isolate strains of viruses of three different species, i.e., human, avian and swine. These isolate strains have been registered from 1958 to September 2009, while the datasets have been generated as a view on Influence Research Database hosted at the NIAID BioHealthBase BRC[1] and contain 3221 isolate strains for human, 1119 isolate strains for swine, and 757 isolate strains for avian.

Experiments are performed to study the effect of the thresholds w, δ, $minGR$ and $minREP$ on the discovered periodic changes and emerging patterns. The parameter $minSUP$ is fixed to 0.1. In this case study, the time-points correspond to years, while the number of the ranges produced by the discretization function is fixed to 5. Statistics on the results are collected in Table 1.

In Table 1(a), we have the number of PCs and EPs when tuning w (δ=20, $minGR = 2$, $minREP = 3$). By increasing the width of the time-windows, the overall number of the time-windows decreases, which results in a shorter succession of pairs of time-windows where finding EPs and PCs. This explains the decrease of the number of EPs and PCs for swine and human. We have a different behavior for the phenotypes of avian, where the number of EPs and PCs increases. Indeed, the use of wider time-windows (w=10 and 15 against w=5 and 7) leads to collect greater sets phenotypes having likely higher changeability. In this case, it seems that phenotype change concerns longer periods.

In Table 1(b), we have the results when tuning δ (w=5, $minGR = 2$, $minREP = 3$). As expected, higher values of δ allow us to detect a more numerous set of PCs, which comprises both the replications of EPs which are closer and the replications of EPs that are distant. Whilst, when δ is 10, we capture only the PCs that cover at most ten years. The threshold δ does not affect the number of EPs since it operates after the extraction of the EPs. In Table 1(c), we have the results when tuning $minGR$ (w=5, δ=20, $minREP = 3$). We observe that $minGR$ has great effect on the number of PCs and on the number of EPs. Indeed, at high values of $minGR$, the algorithm is required to detect the strongest changes of support of the patterns, which leads to extract only the EPs with the higher values of growth-rate. This explains the decrease of the number of PCs. The threshold $minREP$ has no effect on the EPs since it acts on the PCs only (Table 1(d), w=5, δ=20, $minGR = 2$). As expected, higher values of $minREP$ lead to exclude the EPs that have a low number of replications. This means that the algorithm works on a smaller set of EPs, with the result to have a lower number of PCs. In particular, when $minREP$ is 6, we have no PC that includes EPs repeated six times and distant at most 20 years.

---

[1] http://www.fludb.org/brc/home.do?decorator=influenza.

In the following, we report an example of periodic change discovered by the proposed algorithm from the human dataset with $w=10$, $\delta=20$, $minGR = 2$, $minREP=3$:

phenotype(P), clinical_condition(P,E), is_a(E,increased_Virulence), dependent_by(E,M1), is_a(M1,mutation_A199S), mutation_of(M1,T), is_a(T,protein_PB2), dependent_by(E,M2), is_a(M2,mutation_A661T), mutation_of(M2,T), dependent_by(E,M3), is_a(M3,mutation_K702R), mutation_of(M3,T).

Here, T : $\langle([1958; 1967], [1968; 1977]), ([1968; 1977], [1978; 1987]), ([1988; 1997], [1998; 2009])\rangle$, $\psi=[2;3,5]$. The periodic change concerns the clinical condition 'increased_Virulence' with the mutation 'A199S' on the protein 'PB2' and the mutation 'A661T' and 'K702R'. Indeed, the frequency of this pattern increases three times by a factor included in the range [2;3,5]. This happens between the time-windows [1958;1967] and [1968;1977], [1968;1977] and [1978;1987], [1988;1997] and [1998;2009].

Table 1. Total number of the periodic changes and emerging patterns discovered on the three species when tuning the width of the time-window w (a), the maximum admissible distance between consecutive pairs of time-windows δ (b), minimum threshold of growth-rate $minGR$ (c) and minimum threshold of repetitions $minREP$ (d). For each cell, we have reported the statistics as number of PCs–number of the EPs.

	w(years)					δ(years)			
	5	7	10	15		10	15	20	25
swine	11–126	11–126	8–63	2–18	swine	0–126	6–126	11–126	15–126
human	20–176	20–176	10–69	2–44	human	14–176	18–176	20–176	22–176
avian	1–4	1–4	2–10	2–10	avian	0–4	0–4	1–4	1–4

(a)　　　　　　　　　　　　　　　(b)

	$minGR$					$minREP$			
	2	4	6	8		3	4	5	6
swine	11–126	7–68	0–2	0–0	swine	11–126	6–126	2–126	0–126
human	20–176	6–61	0–4	0–0	human	20–176	7–176	3–176	0–176
avian	1–4	0–2	0–2	0–0	avian	0–6	0–2	0–2	0–2

(c)　　　　　　　　　　　　　　　(d)

Meteorology. Meteorological changes and climate cycles have been a discussion topic for years and, in the last decades, they are considered to synthesize predictive models for weather forecasting. They are associated with any recurring cyclical fluctuations within global or regional climate. These fluctuations in atmospheric temperature, sea surface temperature, precipitation or other parameters can be quasi-periodic, often occurring in cycles which cover years, decades or even centuries. Therefore, studying the periodicity of the changes at the level of weather parameters (temperature, air pressure, moisture and wind direction) can be crucial for predicting how the present state of the weather will change.

The dataset was generated from the data bank NOAA [15] by collecting the measurements recorded every day from January 1980 to December 1999 by 270 geographically distributed sensors. The distribution of the sensors delimits a specific geographic area. Totally, we have 83220 daily measurements from which we build data for 228 time-points (the total number of months). For each time-point, the values of the parameters of each sensor are the mean of daily values measured in a month. Then, these are processed by means of an equal-frequency discretization function, in order to assign numeric ranges to the parameters.

Experiments are performed by tuning the thresholds w, δ, $minGR$ and $minREP$ and keeping $minSUP$ fixed to 0.1. The number of the ranges produced by the discretization function for the lists $GRlist$ fixed to 5. Statistics on the results are collected in Table 2.

We can see that they basically confirm the observations drawn for the scenario Virology. Table 2(a) reports the number of PCs and EPs when tuning w ($\delta=72$, $minGR = 2$, $minREP = 3$). We see that the larger the width the lower the number of PCs.

Table 2(b) reports the results when tuning δ ($w=3, minGR=2, minREP=3$). By increasing the maximum admissible distance between consecutive repetitions, we obtain a greater set of PCs. As expected, δ has no effect on the number of EPs.

Table 2(c) reports the response of the algorithm when increasing $minGR$ ($w=3$, $\delta=72$, $minREP = 3$). We observe that there are no PCs with particularly strong changes. Indeed, no change with a numeric growth-rate greater than 6 has been repeated.

As expected, the threshold $minREP$ does affect the PCs but does not the EPs. Indeed, when we require changes with a relatively high number of repetitions (e.g., $minREP=24$), we obtain a smaller set of PCs compared to the set of PCs obtained at $minREP=3$.

Table 2. Total number of the periodic changes and emerging patterns discovered when tuning the width of the time-window w (a), the maximum admissible distance between consecutive pairs of time-windows δ (b), minimum threshold of growth-rate $minGR$ (c) and minimum threshold of repetitions $minREP$ (d). In each cell, we have reported the statistics as number of PCs–number of the EPs.

w(months)				δ(months)			
3	6	12	24	12	24	36	72
179–1220	119–678	44–321	28–140	12–678	76–678	154–678	179–678
	(a)				(b)		

$minGR$				$minREP$			
2	4	6	8	3	6	12	24
179–768	121–421	0–17	0–17	179–678	159–678	55–678	3–678
	(c)				(d)		

5 Related Works

The studies on the periodicity have been concentrated on the identification of the occurrences of static behaviors over data sequence. Most of proposed methods follow the frequent pattern mining framework and can be categorized mainly on the basis of the notion of periodicity. The periodicity is asynchronous when the occurrences are not always regular [7], while, we have a partial periodicity [2] when the repetitions do not exhibit always the same elements. Other works have been focused on the periodicity in complex data. For instance, in [8], the authors report an algorithm to find a minimal set of periodically recurring subgraphs that capture all periodic behavior in a dynamic network. They rely on the concepts of closed subgraphs and Occam's razor. In [9], the authors investigated the problem of identifying from trajectory data repeating activities at certain locations with regular time intervals. They proposed a two-stage algorithm able to first localize frequently visited regions and then detects the periods by using the Fourier transform. However, at our knowledge, we cannot enumerate attempts on the study of the periodicity of changing behaviors, despite to the recent interest in change mining problems. In this sense, two main research lines can be distinguished. In the first line, we find methods to identify changes in the global properties of the complex data. In [1] the authors describe a hierarchical clustering technique to identify eras of evolution of a dynamic network. In the second line, we find methods to characterize evolutions of local properties. For instance, in [13], we proposed an algorithm for capturing variations exhibited from the patterns discovered from data network over time. Two different notions of changes were formalized, namely change patterns and change chains, the first one denotes the evolution of the network from a time-period to the next one, the second one denotes the whole evolution over the temporal axis.

6   Conclusions

In this paper we investigated two different aspects of the analysis of complex dynamic data, namely the discovery of changes manifested over time and identification of those changes that repeat over time with a certain periodicity. The proposed method relies on a frequent pattern mining framework which enable us to *(i)* capture the changes with variations of the frequency of frequent patterns and *(ii)* determine the periodicity of the changes with regularly recurring variations of the frequency. At this end, we extended the classical notion of emerging patterns and formalized the new notion of periodic change. A periodic change is an emerging pattern that meets the constraint of periodicity set as input parameter, that is, the maximum period with which the emerging pattern recurs is known. However, often changes can repeat with unexpected periods, hence using a known value of periodicity could miss some patterns. This is the main future direction we intend to upgrade the method. The experiments highlights the viability of the proposed method to real-world problems where the study of variations repeated over time is an actual problem.

Acknowledgements. The authors would like to acknowledge the support of the European Commission through the project MAESTRA - Learning from Massive, Incompletely annotated, and Structured Data (Grant number ICT-2013-612944).

References

1. Berlingerio, M., Coscia, M., Giannotti, F., Monreale, A., Pedreschi, D.: Evolving networks: Eras and turning points. Intell. Data Anal. **17**(1), 27–48 (2013)
2. Chen, S., Huang, T.C., Lin, Z.: New and efficient knowledge discovery of partial periodic patterns with multiple minimum supports. J. Syst. Softw. **84**(10), 1638–1651 (2011)
3. Dong, G., Li, J.: Efficient mining of emerging patterns: Discovering trends and differences. In: Proceedings of the Fifth ACM SIGKDD International Conference on Knowledge Discovery and Data Mining, pp. 43–52 (1999)
4. Ferlez, J., Faloutsos, C., Leskovec, J., Mladenic, D., Grobelnik, M.: Monitoring network evolution using MDL. In: Alonso, G., Blakeley, J.A., Chen, A.L.P. (eds.) Proceedings of the 24th International Conference on Data Engineering, ICDE 2008, April 7–12, 2008, Cancún, México, pp. 1328–1330. IEEE (2008)
5. Furuse, Y., Suzuki, A., Kamigaki, T., Oshitani, H.: Evolution of the m gene of the influenza a virus in different host species: large-scale sequence analysis. Virol. J. **6**(67) (2009)
6. Han, J., Dong, G., Yin, Y.: Efficient mining of partial periodic patterns in time series database. In: Proceedings of the 15th International Conference on Data Engineering, Sydney, Austrialia, March 23–26, 1999, pp. 106–115 (1999)
7. Huang, K., Chang, C.: SMCA: a general model for mining asynchronous periodic patterns in temporal databases. IEEE Trans. Knowl. Data Eng. **17**(6), 774–785 (2005)
8. Lahiri, M., Berger-Wolf, T.Y.: Periodic subgraph mining in dynamic networks. Knowl. Inf. Syst. **24**(3), 467–497 (2010)
9. Li, Z., Ding, B., Han, J., Kays, R., Nye, P.: Mining periodic behaviors for moving objects. In: Rao, B., Krishnapuram, B., Tomkins, A., Yang, Q. (eds.) Proceedings of the 16th ACM SIGKDD International Conference on Knowledge Discovery and Data Mining, Washington, DC, USA, July 25–28, 2010, pp. 1099–1108. ACM (2010)
10. Loglisci, C.: Time-based discovery in biomedical literature: mining temporal links. IJDATS **5**(2), 148–174 (2013)
11. Loglisci, C., Balech, B., Malerba, D.: Discovering variability patterns for change detection in complex phenotype data. In: Esposito, F., Pivert, O., Hacid, M.-S., Rás, Z.W., Ferilli, S. (eds.) ISMIS 2015. LNCS, vol. 9384, pp. 9–18. Springer, Heidelberg (2015). doi:10.1007/978-3-319-25252-0_2
12. Loglisci, C., Ceci, M., Malerba, D.: Discovering evolution chains in dynamic networks. In: New Frontiers in Mining Complex Patterns - First International Workshop, NFMCP 2012, Held in Conjunction with ECML/PKDD 2012, Bristol, UK, September 24, 2012, Revised Selected Papers, pp. 185–199 (2012)
13. Loglisci, C., Ceci, M., Malerba, D.: Relational mining for discovering changes in evolving networks. Neurocomputing, **150**, Part A: 265–288 (2015)
14. Plotkin, G.D.: A note on inductive generalization. Mach. Intell. **5**, 153–163 (1970)
15. Simons, R.A.: Erddap - the environmental research division's data access program (2011). http://coastwatch.pfeg.noaa.gov/erddap. *Pacific Grove, CA:NOAA/NMFS/SWFSC/ERD*

Classification

The Usefulness of Roughly Balanced Bagging for Complex and High-Dimensional Imbalanced Data

Mateusz Lango and Jerzy Stefanowski[✉]

Institute of Computing Science, Poznań University of Technology,
60-965 Poznań, Poland
{Mateusz.Lango,Jerzy.Stefanowski}@cs.put.poznan.pl

Abstract. Under-sampling generalizations of bagging ensembles improve classification of imbalanced data better than other ensembles. Roughly Balanced Bagging is the most accurate among them. In this paper, we experimentally study its properties that may influence its good performance. Results of experiments show that it can be constructed with a small number of component classifiers. However, they are less diversified than components of the standard bagging. Moreover, its good performance comes from its ability to recognize unsafe types of minority examples better than other ensembles. We also present how to improve its performance by integrating bootstrap sampling with the random selection of attributes.

Keywords: Class imbalance · Roughly Balanced Bagging · Types of minority examples · High-dimensional data · Random Subspace method

1   Introduction

Learning classifiers from imbalanced data still reveals research challenges. However, difficulties are not caused by the unbalanced class cardinalities only. Deterioration of classification performance arises when other data difficulty factors occur together with the class imbalance ratio, such as decomposition of the minority class into rare sub-concepts, too extensive overlapping of decision classes or presence of minority examples inside the majority class regions [10,14].

Several methods have been introduced to deal with imbalanced data; for their review see, e.g., [6]. They are usually categorized into data level and algorithm level approaches. Methods of the first category pre-processes the original data to change their distribution into a more appropriate one (usually more balanced). The other category involves specific solutions to modify the learning algorithm. Modifications of ensembles are also among them; see their review in [5,13]. Most of them are usually modifications of bagging or boosting. They either employ pre-processing methods before learning component classifiers or embed the cost-sensitive framework in the learning process. However, there is still a lack of more comprehensive studies of their properties.

© Springer International Publishing Switzerland 2016
M. Ceci et al. (Eds.): NFMCP 2015, LNAI 9607, pp. 93–107, 2016.
DOI: 10.1007/978-3-319-39315-5_7

Studies [5, 11] have shown that extensions of bagging work better than generalizations of boosting and other, more complex solutions. Moreover, experiments from [1] have shown that under-sampling modifications of bagging are significantly better than over-sampling alternatives. The recent studies [1, 2, 11] have demonstrated that the under-bagging called Roughly Balanced Bagging (RBBag) [7] has achieved the best results comparing to other extensions of bagging as well as modified boosting ensembles [11].

The key idea behind Roughly Balanced Bagging is a random under-sampling before generating component classifiers, which reduces the presence of the majority class examples inside each bootstrap sample. Although this ensemble has been successfully used in several studies, there are not enough attempts to check which of its properties are the most crucial for improving classification of complex imbalanced data. In our opinion, they should be examined more precisely.

Thus, the main aim of this paper is to experimentally study the following issues: (1) the most influential aspects of constructing RBBag and its main properties (with respect to bootstrap construction, deciding on the number of component classifiers, their diversity, methods for aggregating predictions); (2) abilities of this ensemble to deal with different types of difficult distributions of the minority class; (3) directions for its further extension and improvements.

The other aim is to present a new extension of Roughly Balanced Bagging which integrates its specific bootstrap sampling with a random selection of attributes. This should improve classification of highly imbalanced data in such domains as image recognition, fraud detection and genetic data analysis [17].

2   Related Works

For the reviews of ensembles dedicated to class imbalance consult [5, 6, 13]. These authors distinguish mainly *cost-sensitive* approaches vs. integrations with *data pre-processing*. Below we briefly present under-bagging proposals only, which are most relevant to our study.

Breiman's bagging samples a number of subsets from the training set, builds multiple base classifiers and aggregates their predictions to make a final decision. *Bootstraps* are generated by uniform random sampling with replacement of instances from the original training set (usually keeping the size of the original set). However, as this sampling is performed on all data elements, regardless their class labels (majority or minority), the imbalanced class distribution will be hold in each bootstrap and the ensemble will fail to sufficiently classify the minority class. Most of current proposals overcome this drawback by applying pre-processing techniques to each bootstrap sample, which change the balance between classes – usually leading to the same, or similar, cardinality of the minority and majority classes. For instance, the over-sampling methods typically replicate the minority class data (either by random sampling or generating synthetic examples) to balance bootstraps.

In *under-bagging* the number of the majority class examples in each bootstrap is reduced to the cardinality of the minority class (N_{min}). In the simplest proposals, as *Exactly Balanced Bagging* [3], the entire minority class is just copied to the bootstrap and then combined with the randomly chosen subset of the majority class to exactly balance the cardinality between classes.

While such under-bagging strategies seem to be intuitive and work efficiently in some studies, Hido et al. [7] have claimed that they do not truly reflect the philosophy of bagging and could be still improved. In the original bagging the class distribution of each sampled subset varies according to the binomial distribution while in the above under-bagging strategy each subset has the same class ratio as the desired balanced distribution. In *Roughly Balanced Bagging* (RBBag) the numbers of instances for both classes are determined in a different way by equalizing the sampling probability for each class. The number of minority examples (S_{min}) in each bootstrap is set to the size of the minority class N_{min} in the original data. In contrast, the number of majority examples is decided probabilistically according to the negative binomial distribution, whose parameters are the number of minority examples (N_{min}) and the probability of success equal to 0.5. In this approach only the size of the majority examples (S_{maj}) varies, and the number of examples in the minority class is kept constant since it is small. Finally, component classifiers are induced by the same learning algorithm from each i bootstrap sample ($S^i_{min} \cup S^i_{maj}$) and their predictions form the final decision with the equal weight majority voting.

Hido et al. compared RBBag with several algorithms showing that it was better on G-mean and AUC measures [7]. The study [11] demonstrated that under-bagging ensembles, including RBBag, significantly outperformed best extensions of boosting and the difference was more significant when data were more noisy. Then, the comparative experiments from [1] showed that under-bagging ensembles, as Exactly and Roughly Balanced bagging, were significantly better than all main over-sampling extensions of bagging (either using random over-sampling or SMOTE) with respect to all evaluated measures. Roughly Balanced Bagging was also slightly better than Exactly Balanced one. This is why we have chosen RBBag as the best performing specialized ensemble for this paper. The results of [1] also support using sampling with replacement in RBBag – so, we will also use it in further experiments. Recall the introductory motivation that there have not been so many attempts either to experimentally examine properties of this ensemble or to more theoretically explain why and when it should outperform other methods. Only the work [18] provides a probabilistic theory of imbalance and its reference to under-sampling ensembles. Another recent work [4] proposes a theoretical analysis specifying under which conditions under-sampling could be effective in pre-processing for a single classifier.

3   Studying the Role of Components in Roughly Balanced Bagging

The first part of experiments aims at studying the following basic properties of constructing Roughly Balanced Bagging, which have not been studied in the literature yet: (1) Using different learning algorithms to build component classifiers; (2) The influence of the number of component classifiers on the final performance; (3) The role of diversity of component classifier's predictions.

We extend the previous implementation of RBBag done by L. Idkowiak for the WEKA framework [1]. We choose 24 UCI datasets which have been used in the most related experimental studies [2,11,14,15]. Moreover, these datasets represent different difficulty factors referring to distributions of the minority class. Following [15] we distinguish between easier (safe) distributions and more difficult ones, including borderline, rare or outlier examples. For a more detailed explanation of these types of distributions see also Sect. 4. In Table 1 we present characteristics of the chosen datasets with respect of these properties and the global imbalance ratio (IR).

Table 1. Datasets characteristics.

Dataset	# examples	# attrib.	IR	Difficulty type
breast-w	699	9	1.90	safe
vehicle	846	18	3.25	safe
new-thyroid	215	5	5.14	safe
abdominal-pain	723	13	2.58	safe
acl	140	6	2.50	safe
scrotal-pain	201	13	2.41	safe/borderline
car	1728	6	24.04	safe/borderline
ionosphere	351	34	1.79	safe/borderline
pima	768	8	1.87	borderline
bupa	345	6	1.38	borderline
hepatitis	155	19	3.84	borderline
credit-g	1000	20	2.33	borderline
haberman	306	4	2.78	borderline
ecoli	336	7	8.60	borderline/rare
cmc	1473	9	3.42	borderline/rare
transfusion	748	4	3.20	rare
yeast	1484	8	28.10	rare
solar-flareF	1066	12	23.79	rare
postoperative	90	8	2.75	rare
cleveland	303	13	7.66	rare
hsv	122	11	7.71	rare
breast-cancer	286	9	2.36	outlier
abalone	4177	8	11.47	outlier
balance-scale	625	4	11.76	outlier

The performance of ensembles is measured using: *sensitivity* of the minority class (the minority class accuracy), its *specificity* (the accuracy of majority classes), their aggregation to the *geometric mean* (G-mean) and *F-measure*. For their definitions see, e.g., [6]. We have chosen these point measures instead of AUC, as the most of considered learning algorithms produce deterministic outputs. These measures are estimated with the stratified 10-fold cross-validation repeated several times to reduce the variance.

3.1 Choosing Algorithms to Learn Component Classifiers

The related works show that Roughly Balanced Bagging as well as other undersampling extensions of bagging are usually constructed with decision trees. In this study, we check whether classification performance of this ensemble may depend on using other learning algorithms. Besides J48 unpruned tree we considered Naive Bayes tree, rule algorithms – Ripper and PART, Naive Bayes classifiers and SVM – all available in WEKA. The RBBag ensemble was constructed with different numbers (30, 50 and 70) of component classifiers.

Here, we summarize the results of the Friedman test only. For all considered evaluation measures we were unable to reject the null hypothesis on equal performance of all versions of RBBag. For instance, average ranks in the Friedman test for G-mean (the smaller, the better) were the following: SVM 4.1; Ripper 4.12; NBTree 4.4; J48tree/PART 4.5; NB 4.8. Quite similar rankings were obtained for other measures. All these results did not show significant differences of using any of these algorithms inside RBBag.

However, for each single algorithm RBBag was significantly better than its standard bagging equivalent (according to the paired Wilcoxon test).

3.2 The Influence of the Number of Component Classifiers

Related works showed that RBBag was constructed with rather a high number of component classifiers. Hido et al. [7] tested it with 100 C4.5 trees. In the study [11] authors applied a dozen of components. Then, 30, 50 or 70 trees were considered in [1]. Thus, we have decided to examine more systemic other (also smaller) sizes of this ensemble and its influence on the final performance. We stayed with learning components with J48 unpruned trees, and for each dataset we constructed a series of Roughly Balanced Bagging ensembles increasing its size one by one - so the number of component classifiers changed from 2 trees up to 100 ones. We present the changes of G-mean values for all datasets in Fig. 1.

For almost all considered datasets increasing the number of component classifiers improves the evaluation measures up to the certain size of the ensemble. Then, values of measures stay at a stable level or slightly vary around a certain level. Note that the RBBag ensemble achieves good performance for a relatively small number of component classifiers. For most datasets, the stable highest value of G-mean is observed approximately between 10 and 15 trees. In case of the sensitivity or F-measure we noticed similar tendencies.

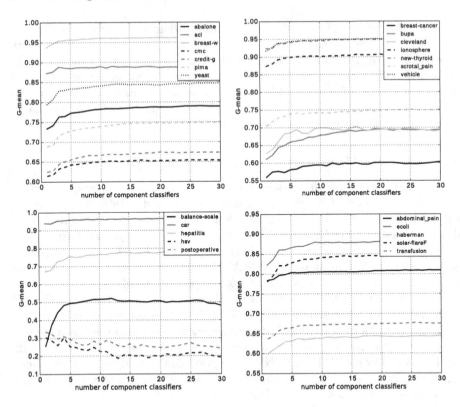

Fig. 1. G-mean vs. a number of component classifiers in RBBag.

Changing the number of components gives a slightly different effect on hsv and postoperative datasets. Observe that augmenting the number of components on these datasets causes a slight decrease of sensitivity and G-mean values, instead of increasing them. Then, values start to fluctuate around certain levels. However, both datasets are the smallest ones as well as the distributions of the minority class are the most sparse and most difficult ones [15].

Moreover, we decided to examine confidence of the final decision of RBBag. We refer to a *margin* of the ensemble prediction. For standard ensembles, it is defined as a difference between the number of votes of components for the most often predicted class label and the number of votes for the second predicted label. Here, we modified it as: $marg = (n_{cor} - n_{incor})/n_{cptclas}$, where n_{cor} is the number of votes for the correct class, n_{incor} is the number of votes for the incorrect class and $n_{cptclas}$ is the number of component classifiers in the ensemble. The higher absolute value of $marg$ is interpreted as high confidence while values closer to 0 indicate uncertainty in making a final decision for a classified instance. In Fig. 2 we present a representative trend of changes of the relative margin with the size of RBBag for ecoli and cmc data. For many other datasets the trend line of the margin also stabilizes after a certain size (Note the resolution of the margin scale

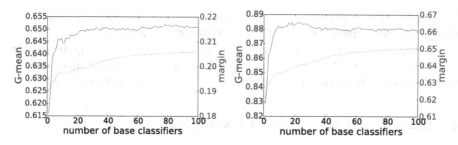

Fig. 2. G-mean and margin vs. a number of component classifiers in RBBag for cmc (*left*) and ecoli (*right*) datasets.

is more detailed than G-mean, so margin values achieve a satisfactory level also quite fast). We can conclude that the good performance of Roughly Balanced Bagging comes from rather a small number of component classifiers.

3.3 Diversity of Component Classifiers

The final accuracy of ensembles may be also related to their diversity - which is usually understood as the degree to which component classifiers make different decisions on one problem (in particular, if they do not make the same wrong decisions). Although, such an intuition behind constructing diverse component classifiers is present in many solutions, research concerns the total accuracy perspective [12]. It is still not clear how diversity affects classification performance especially on minority classes. The only work on ensembles dedicated for imbalance data [19] does not provide a clear conclusion. Its authors empirically studied diversity of specialized over-sampling ensembles and noticed that larger diversity improved recognition of the minority class, but at the cost of deteriorating the majority classes. However, nobody has analysed diversity of RBBag.

To evaluate diversity we calculated the *disagreement measure* [12]. For a pair of classifiers it is defined as a ratio of the number of examples on which both classifiers make different predictions to the number of all classified examples. This measure is calculated for each pair of component classifiers. Then, the global, averaged disagreement D of an ensemble is averaged over all pairs of classifiers. The larger its value is, the more diverse classifiers are [12]. We calculated the global average disagreement D for predictions in both classes and also for the minority class only (denoted as D_{min}). These values are presented in Table 4 - two first columns for RBBag ensemble and next columns refer to its extension discussed in Sect. 5 – both ensembles were constructed with 30 component J48 trees. As this table concerns further extension of RBBag for a higher number of attributes, the list of datasets is reduced.

Notice that values of disagreement measures are relatively low. For nearly all datasets they are between 0.1 and 0.3. The small diversity concerns both class predictions (D) and minority class (D_{min}), although D_{min} is usually lower than D. Similar low values occurred for all remaining datasets, not included in

Table 4. We also checked that changing the number of component classifiers in RBBag did not influence values of the disagreement measure.

To sum up, the high accuracy of RBBag may not be directly related to its higher diversity. We have also analysed predictions of particular pairs of classifiers and noticed that they quite often make the same correct decisions.

4   Influence of the Type of Examples

According to [14,15] the data difficulty factors concerning distributions of imbalanced classes can be modeled by the following types of examples: *safe examples* (located in the homogeneous regions populated by examples from one class only); *borderline* (placed close to the decision boundary between classes); *rare examples* (isolated groups of few examples located deeper inside the opposite class), or *outliers*. Following the method introduced in [14] the type of example can be identified by analysing class labels of the k-nearest neighbours of this example. For instance, if $k = 5$, the type of the example is assigned in the following way [14,15]: 5:0 or 4:1 – an example is labelled as safe example; 3:2 or 2:3 – borderline example; 1:4 – labelled as rare example; 0:5 – example is labelled as an outlier. This rule can be generalized for higher k values, however, results of recent experiments [15] show that they lead to a similar categorization of considered datasets. Therefore, in the following study we stay with $k = 5$.

Similarly to [14,15] we observed that the most of datasets considered in this paper contain rather a small number of safe examples from the minority class. The exceptions are two datasets composed of many safe examples: new-thyroid and car. Many datasets such as cleveland, balance-scale or solar-flare do not contain any safe examples but many outliers and rare cases. The similar analysis of the majority class shows that the datasets contain mostly safe types of majority examples. Recalling recent experiments from [2] we add that differences between performance of various generalizations of bagging are smaller for datasets where safe minority examples dominate inside the distribution. On the other hand, RBBag stronger outperforms other generalizations if datasets contain many unsafe minority examples.

In the current experiments we identified a type of the testing example and recorded whether it was correctly classified or not. Additionally, we refer types of examples in both (minority and majority) classes to the relative margins of the RBBag predictions (these are presented as histograms of numbers of testing examples with a given value of the margins). In Figs. 3 and 4 we present a representative results of RBBag and the standard bagging for cleveland dataset. Histograms for other datasets present similar observations.

Notice that RBBag quite well recognizes the borderline examples from the minority class. Rare minority examples are more difficult, however, on average RBBag can still recognize many of them. It classifies them much better than the standard bagging. Outliers are the most difficult, but RBBag classifies correctly some of them and again this is the main difference to standard bagging and other its over-sampling extensions evaluated in [2]. The similar tendency is observed

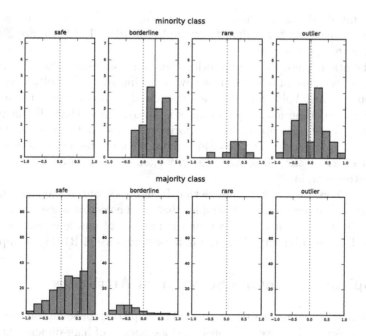

Fig. 3. Histogram of RBBag margins for `cleveland` dataset with respect to a class and a type of example. Blue vertical line shows the value of the margin's median. (Color figure online)

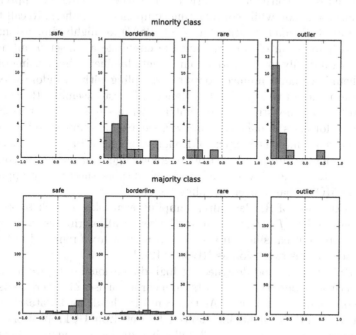

Fig. 4. Histogram of standard bagging margins for `cleveland` dataset with respect to a class and a type of example. Blue vertical line shows the value of the margin's median. (Color figure online)

for other unsafe datasets which are not visualized due to page limits. If the dataset contains some safe minority examples, nearly all of them are correctly classified with high margins.

On the other hand, for the majority class, one can notice that RBBag correctly classifies most of safe examples while facing some difficulties with borderline ones. It also holds for other non-visualized datasets (where the margin's median for borderline majority examples is always worse than the median for borderline minority examples). The majority class does not contain any rare or outlying examples for nearly all considered datasets. For few exceptions as pima, breast-cancer or cmc, these rare majority examples are misclassified with the high negative margin.

In conclusion, we can hypothesize that Roughly Balanced Bagging improves recognition of unsafe minority examples, but at the cost of worse dealing with unsafe majority examples. However, as the number of unsafe examples is relatively small in the majority class, the final performance of RBBag is improved.

5   Applying a Random Selection of Attributes

Although RBBag performs well, it can still be improved for some complex classification problems. Here, we will consider classification of high-dimensional data.

In our proposals we will follow the line of modifying the construction of bootstrap samples. Observations of rather limited diversity of RBBag components have led us to consider inspiration from earlier research on applying random attribute selection while constructing component classifiers. Recall that Ho introduced in [8] *Random Subspace* method (RSM) for highly dimensional data, where in each iteration of constructing the ensemble a subset of all available attributes is randomly drawn and a component classifier is built using only this subset. Then, Breiman combined bootstrap sampling with a random selection of attributes in nodes of trees inside the Random Forest ensemble. Recent experiments of [13] also demonstrated that combing instance re-sampling with Random Forests helps for class imbalance. However, we are more interested in adapting Random Subspace into the context of Roughly Balanced Bagging as it is a classifier independent strategy. To best of our knowledge it has not been considered for RBBag yet. In the only related work [9] authors successfully applied this method to SMOTE based over-sampling bagging.

In our extension of RBBag, after sampling examples to each bootstrap we also randomly select f attributes from the set of all attributes. Subsequently, we train component classifier on a sample from which we removed not selected ones. We denote this extension as RBBag+RSM.

Since RSM is a method designed for high-dimensional data, we have chosen to our experiments only these datasets from earlier phases of experiments, which contain more than 11 attributes. As this condition holds for 9 datasets only, we added 4 new, high-dimensional imbalanced datasets from UCI repository. Finally, in this experiment we examine 13 following datasets: abdominal-pain (13 attributes), cleveland (13), credit-g (20), dermatology (35), hepatitis (19),

Table 2. G-mean for Roughly Balanced Bagging (RBBag) and its modification by a random attribute selection (RBBag+RSM).

Dataset	RBBag			RBBag+RSM		
	30	50	70	30	50	70
abdominal-pain	0.8077	0.8072	0.8062	0.8336	0.8411	0.8358
cleveland	0.7161	0.7247	0.7208	0.6938	0.7197	0.7410
credit-g	0.6735	0.6755	0.6792	0.6930	0.6923	0.7007
dermatology	0.9868	0.9864	0.9873	0.9986	1.0000	1.0000
hepatitis	0.7663	0.7947	0.7920	0.8131	0.8113	0.8029
ionosphere	0.9063	0.9079	0.9098	0.9068	0.9104	0.9152
satimage	0.8727	0.8734	0.8752	0.8677	0.8678	0.8698
scrotal-pain	0.7484	0.7414	0.7455	0.7869	0.7846	0.7884
segment	0.9892	0.9895	0.9896	0.9945	0.9955	0.9953
seismic-bumps	0.6824	0.6945	0.6937	0.7103	0.7153	0.7124
solar-flare	0.8499	0.8511	0.8529	0.8351	0.8437	0.8458
vehicle	0.9525	0.9548	0.9552	0.9590	0.9588	0.9599
vowel	0.9623	0.9604	0.9606	0.9751	0.9766	0.9789

ionosphere (34), satimage (37), scrotal-pain (13), segment (20), seismic-bumps (19), solar-flare (12), vehicle (18) and vowel (14).

We tested it with J48 decision tree (without pruning) and SVM as base classifiers. Following the literature review, we considered setting f parameter to $\lceil \sqrt{F} \rceil$, $\lceil \log_2 F + 1 \rceil$ and $\lceil {}^1/_2 F \rceil$, where F is the total number of attributes in the dataset. Due to space limit we present results only for J48 decision trees and $f = \lceil \sqrt{F} \rceil$, since this parameter setting gives, on average, the highest increments.

Another issue concerns the size of RBBag+RSM. Although bagging can be constructed with a small number of components (for RBBag approx. 15), the survey of literature shows that their number in RSM or Random Forests should be higher as the randomization of attributes increases the variance of bootstrap samples. This is why we will compare RBBag against the RBBag+RSM ensemble with more components: 30, 50 and 70.

The values of G-mean and sensitivity are presented in Tables 2 and 3, respectively. One can notice increases of both measures, in particular RBBag+RSM with more trees. For instance, the increase on sensitivity (abdominal-pain, hepatitis – above 6 %) and G-mean (abdominal-pain, hepatitis, scrotal-pain, seismic-bumps – above 3 %). We performed the paired Wilcoxon test to compare RBBag+RSM against RBBag. With the confidence $\alpha = 0.05$, RBBag+RSM is better on G-mean for 50 ($p = 0.021$) and 70 ($p = 0.013$) trees and nearly for 30 trees ($p = 0.054$). Similar results we obtained for the sensitivity measure.

Table 3. Sensitivity for Roughly Balanced Bagging (RBBag) and its modification by random attribute selection (RBBag+RSM).

Dataset	RBBag			RBBag+RSM		
	30	50	70	30	50	70
abdominal-pain	0.7955	0.7975	0.7925	0.8523	0.8623	0.8563
cleveland	0.7067	0.7175	0.7100	0.6800	0.7117	0.7567
credit-g	0.6610	0.6637	0.6657	0.6493	0.6407	0.6540
dermatology	0.9900	0.9950	0.9950	1.0000	1.0000	1.0000
hepatitis	0.7500	0.7917	0.7950	0.8200	0.8267	0.8267
ionosphere	0.8553	0.8561	0.8593	0.8660	0.8737	0.8796
satimage	0.8690	0.8726	0.8753	0.8738	0.8720	0.8777
scrotal-pain	0.7400	0.7330	0.7360	0.7467	0.7560	0.7453
segment	0.9863	0.9875	0.9875	0.9918	0.9933	0.9930
seismic-bumps	0.6312	0.6547	0.6529	0.6624	0.6629	0.6612
solar-flare	0.8690	0.8705	0.8730	0.8450	0.8670	0.8670
vehicle	0.9688	0.9703	0.9724	0.9990	0.9990	0.9990
vowel	0.9667	0.9667	0.9667	0.9911	0.9911	0.9900

In additional experiments we also observed that RBBag+RSM needs more components than RBBag, e.g. for 15 trees there was no significant differences in values of G-mean ($p = 0.11$). It confirms our expectations and earlier literature opinions saying that while introducing random attribute selection one should use more components than in the standard bagging.

We also analysed the results of specificity to see whether good recognition of the minority class is not achieved at a high cost of majority class accuracy. Surprisingly, for most datasets this measure has actually increased and the highest decrease does not exceed 2 %.

Additionally, we calculated the disagreement measure for all examples (D) and also the minority class (D_{min}). The values presented in Table 4 are calculated for 30 trees. For the reader's convenience we present results together with difference of disagreement between RBBag+RSM and original RBBag.

One can notice that Random Subspace method resulted in an increase of disagreement on almost all data sets (except seismic-bumps). Interestingly, despite a decline of the disagreement measure on this dataset we observed improvements on both G-mean and sensitivity. The further analysis of diversity shows that the highest increase is for safe majority and borderline minority examples.

The analysis of histograms of decision margins for RBBag+RSM shows that it increases the margin on rare and outlier minority examples. Also, more safe and borderline minority instances are classified correctly, although the average of the margin slightly decreases. Due to increased diversity, fewer examples are

Table 4. Disagreement measures, calculated for examples from both classes (D) and from the minority class only (D_{min}), for Roughly Balanced Bagging (RBBag) and its modification by random attribute selection (RBBag+RSM).

Dataset	RBBag		RBBag+RSM		Difference	
	D	D_{min}	D	D_{min}	D	D_{min}
abdominal-pain	0.1564	0.1310	0.2995	0.2580	0.1431	0.1269
cleveland	0.2807	0.2470	0.3506	0.3050	0.0700	0.0581
credit-g	0.2648	0.2279	0.4075	0.3951	0.1427	0.1672
dermatology	0.0211	0.0162	0.1815	0.1384	0.1604	0.1222
hepatitis	0.2476	0.2127	0.3156	0.2915	0.0680	0.0788
ionosphere	0.0733	0.0909	0.1158	0.1650	0.0424	0.0741
satimage	0.1549	0.1160	0.1782	0.1448	0.0233	0.0288
scrotal-pain	0.1871	0.1670	0.3522	0.3139	0.1651	0.1469
segment	0.0168	0.0106	0.0659	0.0293	0.0491	0.0187
seismic-bumps	0.2891	0.2373	0.2470	0.2383	−0.0421	0.0010
solar-flare	0.1062	0.0999	0.2362	0.2395	0.1300	0.1396
vehicle	0.0592	0.0509	0.1461	0.0972	0.0869	0.0463
vowel	0.0461	0.0251	0.2126	0.0825	0.1665	0.0574

classified with a maximum decision margin. RBBag+RSM decreases the margin of safe majority examples, but this does not adversely affect the final prediction.

6 Discussion and Final Remarks

This study attempts to extend knowledge on properties of Roughly Balanced Bagging, which is one of the most accurate ensemble dedicated for class imbalances. Our experiments show that it can be constructed with a relatively small number of component classifiers (approx. 15 ones). It is an interesting observation, as this ensemble may require a heavy under-sampling. One could expect that due to such strong changes inside distributions in bootstrap samples, their variance will be high, and the ensemble should reduce it by applying many components. However, the experimental results have shown that it is not a case. Moreover, this can be an indication to construct this ensemble in an iterative way (starting from a smallest size and stepwise adding a new component while testing it with the extra validation set). According to other experiments the choice of the considered algorithms for learning component classifiers does not influence the final performance of RBBag.

Another discovery is quite low diversity of RBBag. We have also confirmed it by calculating Q-statistics diversity measure [12,19]. Comparing it to earlier results [1] we argue that RBBag is less diversified than over-bagging or SMOTE-based bagging. On the other hand, RBBag is more accurate than these more diversified ensembles. We have also checked that its components are quite accurate and pairs of classifiers often make the same correct decisions. It may open another research on studying the trade-off between accuracy and diversity of ensembles for imbalanced data.

Studying the local recognition of types of classified examples shows that RBBag improves classification of unsafe minority examples. Its power for dealing with borderline, rare and outlying examples distinguishes it from other ensembles. Here, we recall results from [2], which revealed that types of several unsafe minority examples from the original data were changed by RBBag bootstrap sampling into safer ones, which was not a case for other bagging extensions.

In spite of good performance of RBBag, we ask questions about its further extension. In this study, we demonstrate that an integration of a random selection of attributes improves predictions for more dimensional complex data. We also observed that: (1) it increases diversity of component classifiers and (2) the higher number of components significantly improves classification, differently than in original RBBag.

However, yet other directions could concern modifications of bootstrap sampling. In [2] we have already introduced Nearest Balanced Bagging which exploits information about types of minority examples and directs sampling toward the more unsafe examples. Although its experimental results are encouraging (for some datasets even better than RBBag) it generates bootstrap samples containing more minority examples than majority ones. So, it may be still shifted too much to improving sensitivity at the cost of not recognizing too many majority examples. Recall that experiments from Sect. 4 have shown that RBBag also improves recognition of unsafe minority examples while worsening classification of borderline majority examples. Therefore, looking for other modifications of sampling which capture trade-off between choosing examples from both classes could be still undertaken.

Furthermore, a decomposition of classes into sub-concepts [10] could be considered. In [16] authors applied k-means clustering to stratify sampling majority examples inside their modifications of standard bagging. Looking for another semi-supervised clustering to better handle complex boundaries of data distributions could be yet another direction for future research.

Finally, one can also consider other methods for aggregating votes of component classifiers. In on-going experiments we have observed that the dynamic estimation of weights with respect to the neighbourhood of the new-coming instance is more promising that using global weights.

Acknowledgements. The research was supported by NCN grant DEC-2013/11/B/ ST6/00963.

References

1. Błaszczyński, J., Stefanowski, J., Idkowiak, Ł.: Extending bagging for imbalanced data. In: Burduk, R., Jackowski, K., Kurzynski, M., Wozniak, M., Zolnierek, A. (eds.) CORES 2013. AISC, vol. 226, pp. 273–282. Springer, Heidelberg (2013)
2. Błaszczyński, J., Stefanowski, J.: Neighbourhood sampling in bagging for imbalanced data. Neurocomputing **150A**, 184–203 (2015)

3. Chang, E.Y.: Statistical learning for effective visual information retrieval. In: Proceedings of the ICIP 2003, vol. 3, pp. 609–612 (2003)
4. Dal Pozzolo, A., Caelen, O., Bontempi, G.: When is undersampling effective in unbalanced classification tasks? In: Appice, A., et al. (eds.) ECML PKDD 2015. LNCS, vol. 9284, pp. 200–215. Springer, Heidelberg (2015)
5. Galar, M., Fernandez, A., Barrenechea, E., Bustince, H., Herrera, F.: A review on ensembles for the class imbalance problem: bagging-, boosting-, and hybrid-based approaches. IEEE Trans. Syst. Man Cybern. Part C: Appl. Rev. **99**, 1–22 (2011)
6. He, H., Ma, Y. (eds.): Imbalanced Learning: Foundations. Algorithms and Applications, IEEE - Wiley, Hoboken (2013)
7. Hido, S., Kashima, H.: Roughly balanced bagging for imbalance data. Stat. Anal. Data Min. **2**(5–6), 412–426 (2009). Proceedings of the SIAM International Conference on Data Mining, 143–152 (2008)
8. Ho, T.: The random subspace method for constructing decision forests. Pattern Anal. Mach. Intell. **20**(8), 832–844 (1998)
9. Hoens, T.R., Chawla, N.V.: Generating diverse ensembles to counter the problem of class imbalance. In: Zaki, M.J., Yu, J.X., Ravindran, B., Pudi, V. (eds.) PAKDD 2010. LNCS, vol. 6119, pp. 488–499. Springer, Heidelberg (2010)
10. Jo, T., Japkowicz, N.: Class Imbalances versus small disjuncts. ACM SIGKDD Explor. Newslett. **6**(1), 40–49 (2004)
11. Khoshgoftaar, T., Van Hulse, J., Napolitano, A.: Comparing boosting and bagging techniques with noisy and imbalanced data. IEEE Trans. Syst. Man Cybern. Part A **41**(3), 552–568 (2011)
12. Kuncheva, L.: Combining Pattern Classifiers: Methods and Algorithms, 2d edn. Wiley, Hoboken (2014)
13. Liu, A., Zhu, Z.: Ensemble methods for class imbalance learning. In: He, H., Ma, Y. (eds.) Imbalanced Learning: Foundations, Algorithms and Applications, pp. 61–82. Wiley, Hoboken (2013)
14. Napierala, K., Stefanowski, J.: Identification of different types of minority class examples in imbalanced data. In: Corchado, E., Snášel, V., Abraham, A., Woźniak, M., Graña, M., Cho, S.-B. (eds.) HAIS 2012, Part II. LNCS, vol. 7209, pp. 139–150. Springer, Heidelberg (2012)
15. Napierala, K., Stefanowski, J.: Types of minority class examples and their influence on learning classifiers from imbalanced data. J. Intell. Inf. Syst. (2015). doi:10.1007/s10844-015-0368-1
16. Sobhani, P., Viktor, H., Matwin, S.: Learning from imbalanced data using ensemble methods and cluster-based undersampling. In: Appice, A., Ceci, M., Loglisci, C., Manco, G., Masciari, E., Ras, Z.W. (eds.) NFMCP 2014. LNCS, vol. 8983, pp. 69–83. Springer, Heidelberg (2015)
17. Pio, G., Malerba, D., D'Eila, D., Ceci, M.: Integrating microRNA target predictions for the discovery of gene regulatory networks: a semi-supervised ensemble learning approach. BMC Bioinform. **15**(Suppl. 1), S4 (2014)
18. Wallace, B., Small, K., Brodley, C., Trikalinos, T.: Class Imbalance, Redux. In: Proceedings of the 11th IEEE International Conference on Data Mining, pp. 754–763 (2011)
19. Wang, S., Yao, T.: Diversity analysis on imbalanced data sets by using ensemble models. In: Proceedings of the IEEE Symposium Computational Intelligence Data Mining, pp. 324–331 (2009)

Classifying Traces of Event Logs on the Basis of Security Risks

Bettina Fazzinga[1], Sergio Flesca[2], Filippo Furfaro[2(✉)], and Luigi Pontieri[1]

[1] ICAR-CNR, Rende, Italy
{fazzinga,pontieri}@icar.cnr.it
[2] DIMES, University of Calabria, Rende, Italy
{flesca,furfaro}@dimes.unical.it

Abstract. In the context of security risk analysis, we address the problem of classifying log traces describing business process executions. Specifically, on the basis of some (possibly incomplete) knowledge of the process structures and of the patterns representing unsecure behaviors, we classify each trace as instance of some process and/or as potential security breach. This classification is addressed in the challenging setting where each event has not a unique interpretation in terms of the activity that has generated it, but it can correspond to more activities. In our framework, the event/activity mapping is encoded probabilistically, and the models describing the processes and the security breaches are expressed in terms of precedence/causality rules over the activities. Each trace is classified on the basis of the conformance of its possible interpretations, generated by a Monte Carlo mechanism, to the security-breach models and/or the process models. The framework has been experimentally proved to be efficient and effective.

1 Introduction

Despite the adoption of automated control, monitoring, and tracing infrastructures, and of and access/usage control mechanisms, business processes are continuously exposed to security threats (e.g., financial frauds, data leakage, system faults, regulatory noncompliance). In fact, security breaches tend to emerge frequently in real-world processes, and this may severely undermine the achievement of business goals, or severely damage the organization, as witnessed by recent scandals like the 2011 UBS rogue trader scandal. This clearly calls for analyzing the actual behavior of process instances, exploiting the execution traces that are typically registered by process enactment systems. However, as discussed in [5], due the lack of tools tailored at business process logs, current auditing practices often rely on manual inspections and/or on sampling.

This explains the recent interest [4,5,9,13] towards exploiting process mining techniques (such as workflow induction [1], compliance [3] and conformance [15] checking) for carrying out security-oriented analyses over business process logs. However, these techniques found on the assumption that each step in a log

© Springer International Publishing Switzerland 2016
M. Ceci et al. (Eds.): NFMCP 2015, LNAI 9607, pp. 108–124, 2016.
DOI: 10.1007/978-3-319-39315-5_8

trace unambiguously refers to one of the activities that compose the "high-level" process models that security analysts and business users have in mind. Unfortunately, many enactment and tracing systems work at a lower abstraction level: each event in a trace represents the execution of a fine-grain operation, with no clear mapping to a unique "high-level" activity.

In this work, we face the problem of classifying business process traces in the context of security risk analysis: based on some knowledge on the structures of the processes and of the patterns representing undesired/risky behaviors (encoded into *process models* and *security-breach models*, respectively), we aim at classifying each trace as instance of some process and/or as potential security breach. In particular, the problem is addressed in the above-introduced challenging setting: the models feature high-level activities (and encode precedence/causality constraints over the activities, like in declarative process modeling frameworks [2,16]); conversely, each trace is a sequence of low-level events, and can be regarded as the execution of one among many possible activities. Consequently, multiple interpretation may exist for each trace τ, in terms of sequences of activities that might have generated (the sequence of events composing) τ.

Example of the scenario. *Consider an issue management process within a CRM unit, which logically consists of a number of high-level activities, such as: creating a ticket (*Create*), sending a reply to the customer (*Reply*), assigning a ticket to a solver (*Assign*), writing a report (*Report*), contacting the customer through a private channel (*Contact Privately*), and closing a ticket (*Close*). If the process is not implemented over a structured process-aware system (e.g. a* Workflow Management System*), it may happen that the events traced in the process log just represent low-level generic operations, which cannot be mapped univocally to the above activities. For example, different communication-oriented activities (e.g.,* Reply *and* Contact Privately*), may be simply registered in the log as different occurrences of a low-level event (such as "Message sent"), which represents a generic messaging operation.*

In order to perform a security-oriented analysis of the process behavior, one might want to check whether a private communication occurred without the need, e.g., a long time after the closure of the ticket. Obviously, this requires each trace to be properly interpreted, by mapping each event "Message sent" to one of the activities Reply *or* Contact Privately*, and check whether the events mapped to* Contact Privately *were performed long time after the event mapped onto* Close *in the same trace.*

This means that performing a security analysis in the considered setting requires addressing the uncertainty inherent to interpreting traces as sequences of activity executions. In order to deal with this uncertainty, we encode the mapping from events to activities by way of probability distributions, and address the problem of classifying each trace based on the conformance of its possible interpretations to the security-breach models and/or to the process models. In this regard, we address two possible scenarios: the *open world* scenario, where the set of models in input is possibly incomplete (i.e., some traces might have been produced by unknown process models, and may follow unknown kinds of breaches); and the *closed world* scenario, where a model is given for every possible business processes and security breaches. In the latter scenario, each trace is known to be aligned with at least one of the process models, and it can be (probabilistically) assigned to two classes only ("breaches" vs. "non-breaches"). Two similar further classes are to be considered in the open world scenario, in order to encompass the event that the trace does not comply with none of the given process models. Due to the one-to-many mapping from events to activities, estimating the probability that a trace τ belongs to one of these classes would require exploring a possibly very large (combinatorial) number of possible interpretations of τ. Since this may be unfeasible in many real-world scenarios (as shown in the experimentation section), we propose to adopt an ad-hoc Monte Carlo sampling method, allowing to efficiently obtain good estimates of the class-membership probabilities (based on a reduced number of interpretations). The proposed framework has been experimentally validated, and proved to be efficient and effective.

2   Preliminaries

Logs, traces, processes, activities and events. A log is a set of *traces*. Each trace Φ describes a process instance at the abstraction level of basic *events*, each generated by the execution of an activity. That is, an instance w of a *process* W consists of a sequence $a_1, \ldots, a_n$ of *activity* instances; in turn, each activity instance a_i generates an event e_i; hence, the trace Φ describing w consists of the sequence $e_1, \ldots, e_n$. For any event e_i occurring in a trace, we assume that the starting time point of its execution is represented in the log, and denote it as $e_i.t_s$. We also assume that any activity instance a_i "inherits" the starting time point from the generated event e_i, and denote it as $a_i.t_s$.

In the following, we assume given the sets $\mathcal{W}$, $\mathcal{A}$ of (types of) processes and activities, respectively, and the set $\mathcal{E}$ of the events that can occur in the log. We denote the elements of $\mathcal{W}$ and $\mathcal{A}$ with upper-case alphabetical symbols (such as W, A). The instances of processes and activities, as well as the events, will be denoted with lower-case symbols (such as w, a, e).

Mapping events onto activities. Typically, the correspondence between activities and events is *one to many*, that is: (1) for every activity $A \in \mathcal{A}$, all the executions of A generate the same event e; (2) conversely, an occurrence of e in some trace of the log cannot be univocally interpreted as the result of an execution of A, since there can be another activity $B \in \mathcal{A}$ whose execution may have generated e. We assume that the mapping of each event $e \in \mathcal{E}$ onto the activities is probabilistically modeled by a pdf $p_e(A)$, where A is a random variable ranging over $\mathcal{A}$. Basically, $p_e(A)$ encodes the probability that a generic occurrence of the event e has been caused by an execution of activity A. It is worth noting that $p_e(A)$ can be obtained by encoding the knowledge of domain experts, or (as done in our experiments) using a training set of traces from which statistics can be evaluated on how frequently the event e is generated by an instance of A, for each $e \in \mathcal{E}$ and $A \in \mathcal{A}$. The latter option still implies that, for each event e of any trace in the training set, it is known which high-level activity was executed correspondingly to e. Notably, in many process management scenarios (like those considered in our experiments) where most activities are carried out by human executors, one can ask each of these workers to label the events (of the training log) she/he actually performed. This way, the analyst is freed from the burden of such a manual labelling task, at the cost of some very little extra effort for the workers.

In Sect. 5, we will also experimentally investigate the impact on our framework of encoding the probabilistic mapping between events and activities with bigram models.

Interpretations and their probabilities. An activity instance a of A is said to be an *interpretation* of e (or, equivalently, *compatible* with e) if $p_e(A) > 0$ (meaning that e can be reasonably viewed as the result of executing A). Given a trace $\Phi = e_1 \ldots e_n$, we call *interpretation* of Φ a sequence of activity instances $a_1 \ldots a_n$ such that, $\forall i \in [1..n]$, a_i is an interpretation of e_i. The set of interpretations of Φ will be denoted as $\mathcal{I}(\Phi)$.

$p_e(A)$ can be used as the core of a naive mechanism for defining a pdf over $\mathcal{I}(\Phi)$: assuming independence between the events, each $I = a_1 \ldots a_n$ in $\mathcal{I}(\Phi)$ can be assigned $p(I) = \Pi_1^n p_{e_i}(A_i)$ as the probability of being the sequence of events that generated Φ.

Example 1. Let $\mathcal{A} = \{A, B, C\}$, $\mathcal{E} = \{e_1, e_2\}$, and consider the log $\mathcal{L} = \{\Phi\}$, where $\Phi = e_1 e_2$. Assume that $p_{e_1}(A) = p_{e_1}(B) = 0.5$ (meaning that the executions of activities A and B are equi-probable causes of event e_1), and that $p_{e_2}(C) = 1$ (meaning that every occurrence of event e_2 can be generated only by an execution of C). We denote as a, b, c the generic instances of A, B and C, respectively. Given this, trace Φ can be interpreted as either the sequence of activity instances $I_1 = a\,c$ (with probability $p_{e_1}(A) \cdot p_{e_2}(C) = 0.5$) or $I_2 = b\,c$ (with probability $p_{e_1}(B) \cdot p_{e_2}(C) = 0.5$). □

Process models and security-breach models. We assume that some knowledge of the structure of every process $W \in \mathcal{W}$ can be encoded in terms of a set $W.\mathcal{IC}$ of *composition rules*, restricting the sequences of activity instances that are allowed to be executed within W. Basically, a composition rule has one of the following forms: 1) $A \Rightarrow_T B$; 2) $A \Rightarrow_T \neg B$; 3) $A \Leftarrow_T B$; 4) $\neg A \Leftarrow_T B$; where $A, B \in \mathcal{A}$, while T is of the form '$\leq c$', where c is a constant. Herein, $A \Rightarrow_T B$ (resp., $A \Rightarrow_T \neg B$) imposes that, within every instance of W, the execution of any instance a of A must (resp., must not) be *followed* by the execution of an instance b of B such that the width of the interval between $a.t_s$ and $b.t_s$ satisfies T. The semantics of the forms $A \Leftarrow_T B$ and $\neg A \Leftarrow_T B$ is analogous: they have to be read from the right to the left, replacing the word "*followed*" with "*preceded*" in the above definition. Omitting T is the same as specifying $T = $ '$\leq \infty$'. Special cases of the rules of the form $A \Rightarrow B$ and $A \Rightarrow \neg B$ are the rules $true \Rightarrow B$ and $true \Rightarrow \neg B$ meaning that the activity B must (resp., must not) occur within the executions of the process, respectively.

Example 2. Let W be a process with $W.\mathcal{IC} = \{A \Rightarrow \neg B; C \Leftarrow A\}$. Then, the sequence $a\,b\,c$, whose elements are instances of the activities A, B, C, cannot be an instance of W, since it violates both the composition rules. On the contrary, the sequence $c\,a$ conforms to $W.\mathcal{IC}$, thus it can be viewed as an instance of W.   □

We also assume that some knowledge of security risks is encoded in terms of *security-breach models*. A security-breach model is a set of composition rules describing causality and precedence relationships between activity executions that describe risky situations. The composition rules have the same syntax as those used for the process models. In the following, we denote as $\mathcal{SBM}$ the set of security-breach models, and, for each security-breach model $SBM \in \mathcal{SBM}$, we denote the set of composition rules associated with SBM as $SBM.\mathcal{IC}$.

Example 3. Let SBM be a security breach model with $SBM.\mathcal{IC} = \{\neg A \Leftarrow_{<12hr} B\}$, where A is the activity *"Detection of critical trouble"* and B is *"Communication with the customer via private channel"*. The composition rule of SBM means that the case of a customer who has been contacted via a private channel without the existence of a recently detected critical trouble is a security breach that should be looked into.   □

We point out that we assume that process and security-breach models are described in terms of the same forms of composition rules just to simplify the explanation of our framework: in Sect. 7, we will discuss how the framework can be straightforwardly extended to the case that these models are described using different language and/or more expressive variants of these constraints, while possibly considering a richer description of trace events and activities (including, e.g., their end times and executors).

Open world and closed world assumptions. We consider two scenarios, corresponding to different levels of completeness of the knowledge on the structure of the processes and of the patterns describing security breaches. The open world scenario is the case that this knowledge is incomplete: W (resp., $\mathcal{SBM}$) is a (possibly strict) subset of the set of models describing the structure of all the possible processes (resp., security breaches) that can occur in the log. On the contrary, the closed world scenario is the case that all the possible processes and security breaches conform to some model in W and $\mathcal{SBM}$, respectively. For instance, under the open world assumption, the meaning of $W = \{W_1\}$ and $\mathcal{SBM} = \{SBM_1, SBM_2\}$ is: *the only processes and security risks for which a model is known are W_1 and SBM_1, SBM_2, respectively; however, the occurrence of process instances and security breaches conforming to none of these models cannot be excluded.* Under the closed world assumption, there cannot be process instances that do not conform to W_1, and there are no patterns characterizing security breaches that do not conform to either SBM_1 or SBM_2.

3   The Classification Problem and Our Approach for Solving It

The problem addressed in this paper is that of classifying the log traces on the basis of the knowledge of the process and security-breach models. We define the classification problem considering the open and closed world scenarios separately.

Classifying traces under the open world assumption. In this scenario, any sequence of activity instances belongs to exactly one of the following classes (the set of these classes will be denoted as $C(ow)$):

- *Aligned*: sequences conforming to at least one process model $W \in \mathcal{W}$, but to no security breach model $SBM \in \mathcal{SBM}$;
- *Breach*: sequences conforming to at least one $SBM \in \mathcal{SBM}$, but to no $W \in \mathcal{W}$;
- *Aligned&Breach*: sequences conforming to at least one $W \in \mathcal{W}$ and at least one $SBM \in \mathcal{SBM}$;
- *Unknown*: sequences conforming to no $W \in \mathcal{W}$ and no $SBM \in \mathcal{SBM}$.

Classifying a trace $\Phi = e_1 \ldots e_n$ means marking it with the name of the class containing the interpretation $a_1 \ldots a_n$ that generated $e_1 \ldots e_n$. Since Φ has many possible interpretations, where each $I \in \mathcal{I}(\Phi)$ has probability $p(I)$ of being the actual "origin" of Φ, this problem can be re-written under a probabilistic standpoint: for each class $C \in C(ow)$, evaluate the probability $p^{ow}(\Phi, C)$ that the sequence of activities that actually generated Φ is in C. This means evaluating, for each $C \in C(ow)$, the probability:

$$p^{ow}(\Phi, C) = \sum_{I \in (\mathcal{I}(\Phi) \cap C)} p(I). \tag{1}$$

Observe that, in the open world scenario, the independence assumption on which the definitions of $p(I)$ and $p^{ow}(\Phi, C)$ are based is reasonable, since no structure for the processes can be excluded: this backs assuming no correlation between the activities.

Notably, for any Φ, it holds $\sum_{C \in C(ow)} p^{ow}(\Phi, C) = 1$, as this sum represents the cumulated probability of all the possible interpretations that exist for Φ.

Classifying traces under the closed world assumption. The closed world assumption describes the case that the possible structures of the processes and security breaches are exhaustively described by the models in $\mathcal{W}$ and $\mathcal{SBM}$. Hence, only the classes *Aligned* and *Aligned&Breach* must be considered to mark every trace Φ in the log. The set consisting of these two classes will be denoted as $C(cw)$. As regards the probability $p^{cw}(\Phi, C)$ that, under the closed world assumption, the actual interpretation of Φ belongs to the class C (where $C \in C(cw)$), its definition is more complex than its "open-world" counterpart $p^{ow}(\Phi, C)$. We cannot define $p^{cw}(\Phi, C) = p^{ow}(\Phi, C) = \sum_{I \in \mathcal{I}(\Phi) \cap C} p(I)$ since the definition of $p(I)$ relies on the independence assumption, that is no longer valid in this scenario. Indeed, the fact that only the interpretations satisfying some process model must be considered for the classification purposes means that there are correlations between the activity executions: hence, the events cannot be considered independent from one another. A reasonable way to define $p(\Phi, C)$ is that of applying the probabilistic conditioning paradigm, that is a way to revise the pdf that would hold under independence assumption by a-posteriori enforcing some correlations. In our case, this means defining $p^{cw}(\Phi, C) = p^{ow}(\Phi, C | W.\mathcal{IC})$, where the right-hand side of this expression is the result of conditioning the probability $p^{ow}(\Phi, C)$ to the event that the constraints expressed by the composition rules of at least one process model in $\mathcal{W}$ are satisfied. This means that, for each $C \in \mathcal{C}(cw)$:

$$p^{cw}(\Phi, C) = \frac{p^{ow}(\Phi, C)}{p^{ow}(\Phi, Aligned) + p^{ow}(\Phi, Aligned\&Breach)}. \tag{2}$$

An example of classification. The following example shows an example of computation of the probabilities $p^{ow}(\Phi, C)$ (for each $C \in \mathcal{C}(ow)$) and $p^{cw}(\Phi, C)$ (for each $C \in \mathcal{C}(cw)$), and discusses the meaning of the classification.

Example 4. Consider the case described in Example 1 of the trace $\Phi = e_1 e_2$, having two interpretations: $I_1 = a\,c$ and $I_2 = b\,c$, with $p(I_1) = p(I_2) = 0.5$. Assume that $\mathcal{W} = \{W\}$ and $\mathcal{SBM} = \{SBM\}$, where $W.\mathcal{IC} = \{A \Leftarrow C; B \Rightarrow C\}$ and $SBM.\mathcal{IC} = \{B \Leftarrow C\}$. It is easy to see that I_1 conforms to $W.\mathcal{IC}$, but not to $SBM.\mathcal{IC}$. Moreover, I_2 does not conform to $W.\mathcal{IC}$, but conforms to $SBM.\mathcal{IC}$.

Under the open world assumption, the classes to be considered are: $\mathcal{C}(ow) = \{Aligned, Breach, Aligned\&Breach, Unknown\}$. Then, the classification of Φ consists of the following probability assignment: $p^{ow}(\Phi, Aligned) = p(I_1) = 0.5$; $p^{ow}(\Phi, Breach) = p(I_2) = 0.5$; $p^{ow}(\Phi, Aligned\&Breach) = p^{ow}(\Phi, Unknown) = 0$.

On the contrary, under the closed world assumption, the classes to be considered are: $\mathcal{C}(cw) = \{Aligned, Aligned\&Breach\}$, and the classification of Φ is as follows: $p^{cw}(\Phi, Aligned) = p(I_1)/p(I_1) = 1$; $p^{cw}(\Phi, Aligned\&Breach) = 0$.

The classification under the closed world assumption can be read as follows: with probability 1, Φ is consistent with the model of the process, and contains no security breach. Analogously, the classification under the open world assumption can be read as follows: with probability 0.5, the trace Φ is consistent with a known process model and does not contain any known security breach; with probability 0.5, Φ does not conform to any known process model but conforms to a security breach model.   □

3.1   The Challenges of Evaluating a Classification and Our Solution

Efficiently providing the classification of the traces in a log is a hard problem, independently from the fact that the open world (ow) or the closed world (cw) assumption is made. A naive approach for classifying a trace Φ is the following: (1) generate the set $\mathcal{I}(\Phi)$ of all of the interpretations of Φ; (2) for each $I \in \mathcal{I}(\Phi)$, check whether I conforms to some model in $\mathcal{W} \cup \mathcal{SBM}$; (3) depending on the models to which I conforms and on the assumption, mark I with the proper class in $\mathcal{C}(assumption)$; (4) for every class $C \in \mathcal{C}(assumption)$, compute $p^{assumption}(\Phi, C)$ using Eqs. (1) and (2).

Unfortunately, this naive approach is infeasible, as the interpretations to be considered may be too many. For instance, consider a trace Φ of length 40. If, on average, each event has two interpretations, then $|\mathcal{I}(\Phi)| = 2^{40} = 10^{12}$ interpretations for Φ must be considered. As we will show experimentally (see Sect. 5), deciding the class of all these interpretations typically requires very long waits.

In this paper, we investigate the possibility of using a Monte Carlo approach for evaluating the classification of each log trace Φ. Basically, we use a Monte Carlo sampler over $\mathcal{I}(\Phi)$, that randomly generates a new interpretation in $\mathcal{I}(\Phi)$ until the probabilities of the classes (evaluated only on the basis of the sample set generated so far, instead of the whole $\mathcal{I}(\Phi)$) converge to the actual probability values (the convergence is checked according to a given confidence level). We show that our approach is efficient under both the open and closed world assumptions, and feasible even in the cases where the exhaustive approach requires too much time.

4   The Monte Carlo Classification Algorithm

In this section, we describe our Monte-Carlo simulation approach and its implementation (Algorithm 1). Algorithm 1 takes as input the trace Φ, the sets of process and security breach models $\mathcal{W}$ and $\mathcal{SBM}$, the set P of pdfs $p_e(A)$, the assumption under which the classification must be evaluated (that is, either ow or cw), and two parameters defining the desired guarantee on the accuracy of the Monte Carlo estimation: an error level ϵ and a confidence level $1 - \alpha$. The output of Algorithm 1 is an estimate of the classification of Φ under the specified assumption: that is, Algorithm 1 returns an estimate $\tilde{p}^{assumption}(\Phi, C)$ of $p^{assumption}(\Phi, C)$ for each $C \in \mathcal{C}(assumption)$. In particular, it is guaranteed that the actual probability $p^{assumption}(\Phi, C)$ is in the interval $\tilde{p}^{assumption}(\Phi, C) \pm \epsilon$ with confidence level $1 - \alpha$.

In brief, Algorithm 1 samples the set $\mathcal{I}(\Phi)$ of interpretations and determines the class in {$Aligned$, $Breach$, $Aligned\&Breach$, $Unknown$} into which each sample I falls. At the end of the sampling process, for each $c \in \mathcal{C}(assumption)$, it returns as $\tilde{p}^{assumption}(\Phi, C)$ the fraction of the samples belonging to class C. In more detail, Algorithm 1 works as follows. At line 1, it initializes the four variables n_A, n_B, n_{AB} and n_U used to store the number of sampled interpretations falling in the class $Aligned$, $Breach$, $Aligned\&Breach$, and $Unknown$, respectively. The loop from line 2 to line 13 represents the core of the sampling process. At each iteration of this loop, an interpretation I is generated (lines 3–4), by randomly choosing one of the candidates activities for each event of the trace (line 4). The loop from line 6 to line 8 scans the process models in $\mathcal{W}$: if one of the process model of $\mathcal{W}$ is found such that I satisfies all the constraints in it, a boolean variable $foundAlignment$ becomes **true** and the loop terminates. Analogously, the loop from line 9 to line 11 scans the security breach models in $\mathcal{SBM}$

Algorithm 1. The interpretation algorithm
***
Input: A trace Φ, a set $\mathcal{W}$ of process models, a set $\mathcal{SBM}$ of security breach models, a set P containing a pdf $p_e(A)$ for every event $e \in \mathcal{E}$, a parameter *assumption* $\in \{ow, cw\}$, an error level ϵ, a confidence level $1 - \alpha$.

Output: $\tilde{p}^{ow}(\Phi, C)$ for each C in $\{Aligned, Breach, Aligned\&Breach, Unknown\}$, in the case *assumption*=*ow*; $\tilde{p}^{cw}(\Phi, C)$ for each C in $\{Aligned, Aligned\&Breach\}$, otherwise.

1:   $n_A = 0, n_B, n_{AB} = 0, n_U = 0, I = \perp$
2: **repeat**
3:     **for all** $e_i \in \Phi$ **do**
4:       $I[i] = chooseActivity(e_i)$
5:     *foundAlignment*=**false**, *foundBreach*=**false**
6:     **for all** $W \in \mathcal{W}$ **do**
7:       **if** $checkModel(I, W.IC, P)$ **then**
8:         *foundAlignment*=**true**; **break**
9:     **for all** $SBM \in \mathcal{SBM}$ **do**
10:      **if** $checkModel(I, SBM.IC, P)$ **then**
11:        *foundBreach*=**true**; **break**
12:     *Update* n_A, n_B, n_{AB}, n_U *according to foundAlignment and foundBreach*
13: **until** $errorGuarantee(n_A, n_B, n_{AB}, n_U, assumption, \epsilon, 1 - \alpha)$
14: **if** *assumption*=*ow* **then**
15:     $n = n_A + n_B + n_{AB} + n_U$
16:     $\tilde{p}^{ow}(\Phi, Aligned) = \frac{n_A}{n}, \tilde{p}^{ow}(\Phi, Breach) = \frac{n_B}{n}, \tilde{p}^{ow}(\Phi, Aligned\&Breach) = \frac{n_{AB}}{n},$
       $\tilde{p}^{ow}(\Phi, Unknown) = \frac{n_U}{n}$
17:     **return** $\tilde{p}^{ow}(\Phi, Aligned), \tilde{p}^{ow}(\Phi, Breach), \tilde{p}^{ow}(\Phi, Aligned\&Breach), \tilde{p}^{ow}(\Phi, Unknown)$
18: $n = n_A + n_{AB}$
19: $\tilde{p}^{cw}(\Phi, Aligned) = \frac{n_A}{n}, \tilde{p}^{cw}(\Phi, Aligned\&Breach) = \frac{n_{AB}}{n}$
20: **return** $\tilde{p}^{cw}(\Phi, Aligned), \tilde{p}^{cw}(\Phi, Aligned\&Breach)$
***

and early terminates if a model is found such that I satisfies all the constraints in it. In that case, a boolean variable *foundBreach* becomes **true**. After the execution of the two loops scanning the process and the security breach models, respectively, *foundAlignment* and *foundBreach* are used to determine the class containing I and update n_A, n_B, n_{AB} and n_U accordingly (lines 12): if both (resp., none of the) variables *foundAlignment* and *foundBreach* are **true**, I falls into *Aligned&Breach* (resp., *Unknown*), thus n_{AB} (resp., n_U) is incremented. In the case that only *foundAlignment* (resp., *foundBreach*) is **true**, I falls into *Aligned* (resp., *Breach*), thus n_A (resp., n_B) is incremented.

The generation of samples is halted by function *errorGuarantee*, that exploits the Agresti-Coull interval [6] to detect if, for each $C \in \mathcal{C}(assumption)$ the estimate $\tilde{p}^{assumption}(\Phi, C)$ of $p^{assumption}(\Phi, C)$ obtained with the samples collected so far lies in the interval $\tilde{p}^{assumption}(\Phi, C) \pm \epsilon$ with confidence level $1 - \alpha$. We recall that, according to [6], the error of the estimate $\tilde{p}$ of the target probability p obtained after n samples is guaranteed to be at most ϵ with confidence level $1 - \alpha$ if $n > \overline{n} = \frac{z_{1-\alpha/2}^2 \cdot \overline{p} \cdot (1 - \overline{p})}{\epsilon^2} - z_{1-\alpha/2}^2$, where $z_{1-\alpha/2}$ is the $1 - \alpha/2$ quantile of the normal distribution, and $\overline{p} = \frac{n_x + (z_{1-\alpha/2}^2)/2}{n + z_{1-\alpha/2}^2}$, where n_x is the number of successes in the n samples. Hence, since the error guarantee must be provided for every estimate returned by Algorithm 1, function *errorGuarantee* computes $|\mathcal{C}(assumption)|$ values of $\overline{n}$, one for each estimate $\tilde{p}^{assumption}(\Phi, C)$ that has to be returned. Correspondingly, Algorithm 1 halts only if the number of generated samples is equal to or greater than all these values of $\overline{n}$ (line 13).

We point out that the Agresti-Coull interval is applicable to our case, since it is reliable when the underlying distribution is binomial, and the percentage of times that a generated sample (interpretation) falls into a class C follows a binomial. In fact, for any sampled interpretation I, the probability of success (i.e., the probability that I belongs to C) does not influence the probability of success for the other interpretations picked by the sampler.

5 Experimental Validation

Hardware settings and dataset features. All the experiments were done on an Intel i7 CPU with 12 GB RAM running Windows 8.1. We tested our framework over synthetic real-like data, generated according to the guidelines of the administrative units of a service agency (SA). In this scenario, a process instance is a collection of activities performed by the staff of the units in response to customers' requests. Examples of activities are the creation of a new folder, the preparation of new documents and their insertion into a folder, the updating of existing documents, contacting the customer, etc. Folders are of different categories, and folders of the same category follow the same execution scheme, that describes a process. We were given a set $\mathcal{W}$ of 6 processes and their models in terms of precedence relationships (between activities), that were easily encoded into composition rules of the form used in our framework. The scenario is complex enough that the same activities were shared by different processes, and the same event could be generated by different activities. Table 1 reports the parameters used in the data generation.

We were also given the models of 8 different types of security breaches, and each model was translated into a set of composition rules. On average, each process was described by 10 composition rules, and each security breach by 5 composition rules.

Besides the composition rules, the service agency gave us a set of 100 real traces describing different process instances at the abstraction level of events, along with their actual interpretations (that is, the corresponding sequences of activities). We used these traces and their interpretations for generating both the set P of pdfs of the form $p_e(A)$ and the larger dataset used in the experiments. As regards P, it was obtained by extracting statistics on the actual correspondences events/activities occurring in the pairs trace/interpretation given by the service agency. As regards the dataset, starting from each interpretation I, we generated a set $perturb(I)$ of 100 sequences of activity instances, by suitably perturbing I. Specifically, $perturb(I)$ was initially assigned the set consisting of only I, and then the j-th sequence in $perturb(I)$ was obtained from the $(j-1)$-th one by applying one perturbation, randomly chosen among: (a) replacing a randomly chosen interpretation step with a new instance of a randomly chosen activity having the same starting time; (b) switching a randomly chosen interpretation step with the subsequent one; (c) removing a randomly chosen interpretation step; (d) inserting a new instance of a randomly chosen activity into the sequence at a random position (the starting time of the new activity is randomly generated in the interval between the starting times of the previous and the subsequent activities). Before adding the perturbed sequence of activity instances to $perturb(I)$ as the j-th element, we checked its consistency with the models of the processes in W, and we discarded it in the case of inconsistency with every model (in this case, a new perturbation was tried over the $(j-1)$-th interpretation). Finally, once the generation of every $perturb(I)$ was finished, each I' in $perturb(I)$ was translated into the corresponding trace (i.e., sequence of events), and this was put in the dataset. This way, we obtained a dataset consisting of 10^4 traces.

Table 1. Values of the parameters used in the generation

Parameters			
Num. of different processes	6	Num. of different activities	14
Num. of different events	20	Avg. candidate activities per event	~ 2.3
Min,Max activities per event	1, 8	Avg. candidate processes per activity	~ 3
Avg num. of traces per process	1667	Overall num. of traces	10^4

Term of comparison, and open and closed world scenarios. We compared our approach (denoted as MC in the following) with the naive exhaustive approach described in Sect. 3.1 (denoted as EX). The experiments under the open world assumption were performed by making W and $\mathcal{SBM}$ consist of one half of the process and security breach models provided by the service agency, thus simulating the case that W and $\mathcal{SBM}$ do not encode a complete knowledge of the possible processes and security breaches. Obviously, for the closed world assumption, we put into W and $\mathcal{SBM}$ all the process and security breach models provided by the service agency, respectively.

Results on the efficiency. We start with comparing our Monte Carlo based approach with EX in terms of efficiency. In what follows, the two variants of Algorithm 1, corresponding to adopting either the open or the closed world assumption, are referred to as MC-OW and MC-CW, respectively. We make no distinction between the behaviors of the exhaustive approach under the two assumptions since there is no difference in terms of efficiency: under both the open and closed world assumptions, EX performs the same number of iterations (as all the interpretations must be generated and classified according to the models). All the results presented in what follows were obtained by setting $\epsilon = 0.001$ and $1 - \alpha = 95\%$.

Figure 1(a) shows the average execution times of the considered approaches vs. the trace length. Even if our dataset contains longer traces, this diagram reports only the results for the traces whose length is less than 30, since EX required too much time ($\geq$ 20min) to complete the classification over longer traces. The results for EX are represented by 3 distinct curves, obtained as follows. First, the set of traces was partitioned into 3 sets, denoted as $(2.0 - 2.2)$, $(2.2 - 2.4)$, $(2.4 - 2.6)$: a trace Φ of the dataset belongs to the set $(X - Y)$ if the average number $ActPerEv(\Phi)$ of activities that are possible interpretations for a step of Φ belongs to the interval $(X - Y)$. Then, for each $(X - Y)$, the curve $\text{EX}(X - Y)$ represents the average execution time of EX over the traces in $(X - Y)$. This distinction was not made for MC-OW and MC-CW, since, as expected, their execution times turned out to be insensitive to $ActPerEv(\Phi)$. From this diagram, it turns out that execution times for EX grow exponentially with the trace length (in fact, the shape of the curves of EX are linear in the presence of a logarithmic scale on the y-axis), and that also increasing $ActPerEv(\Phi)$ results in slowing down EX.

Figure 1(b) considers also traces longer than 30 steps, and depicts the average running time of MC-OW and MC-CW vs. trace length (for what explained before, EX is not considered in this diagram). The diagram shows that average execution times grow with the trace length (this sensitiveness was hidden in Fig. 1(a) by the use of a log scale).

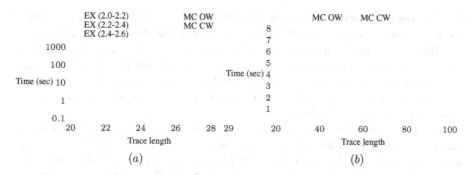

Fig. 1. (a): Execution times of EX, MC-OW and MC-CW vs. trace length over "short" traces; (b): Execution times of MC-OW and MC-CW vs. trace length over all the traces

Results on estimation error of the sampling process. We also analyzed the effectiveness of our Monte Carlo sampling in terms of estimation error of the class membership distributions (the error is evaluated w.r.t. what returned by EX, that considers all the possible interpretations). For each trace Φ, we measured the estimation error of Algorithm 1 as the maximum difference, over all the classes, between the membership probability computed by EX and that estimated by

Algorithm 1. For the same reasons discussed above, only traces shorter than 30 steps were considered. The diagrams in Fig. 2(a, b) report this error under the open and closed world assumption, respectively. In each of these diagrams, three curves are reported, one for each set $(X - Y)$ introduced above. These diagrams show that the average error is, for all the considered trace lengths, lower than the error threshold given as input to Algorithm 1 (we recall that we set $\epsilon = 0.001$, with confidence level $1 - \alpha = 95\%$), and that the error is insensitive to both $ActPerEv(\Phi)$ and the trace length.

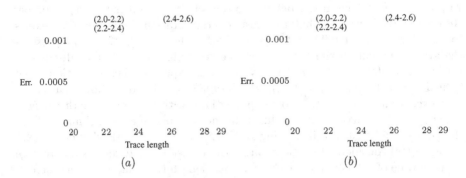

Fig. 2. Average estimation error of Algorithm 1 vs. trace length under the (a) open and the (b) closed world assumption (short traces only)

Results on the accuracy of the classification. In order to assess the utility of our framework, we also validated the accuracy of the classification in terms of its "distance" from the ground truth. Here, for any trace Φ, the ground truth is the class containing the sequence of activities that actually generated Φ. Hence, for a set of traces, the accuracy is measured as the percentage of traces whose actual class coincides with the class that is assigned the highest probability in the classification provided by MC.

As terms of comparison, we considered the following approaches. In the open world scenario, we considered the naive classification algorithm N^{ow}, that computes the most probable interpretation $I^*(\Phi)$ of the input Φ, and decides the class of Φ as the class in $\mathcal{C}(ow)$ containing $I^*(\Phi)$. In the closed world scenario, this technique is not guaranteed to return the most probable valid interpretation, since, in this scenario, only interpretations aligned to $\mathcal{W}$ make sense. Hence, we used a (non-optimal) backtracking greedy approach (denoted as N^{cw}), where the activities with which every step e_i is interpreted are tried in the order implied by $p_{e_i}(A)$.

Moreover, we also investigated the impact of using a bigram model to encode the probabilistic mapping between events and activities, in place of $p_e(A)$. In particular, we considered two alternatives:

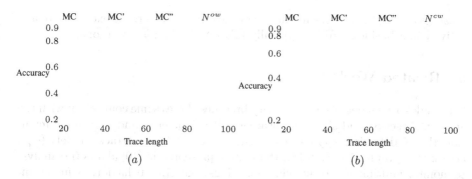

Fig. 3. Accuracy of the classification in the open world (*a*) and the closed world (*b*) scenarios

MC': the use of a pdf $p_e(A|e^{-1})$, representing the probability that an occurrence of e in a trace step is generated by an execution of A, given that the event e is preceded, in the same trace, by e^{-1};

MC": the use of a pdf $p_e(A|A^{-1})$ representing the probability that an occurrence of e in a trace step is generated by an execution of A, given that the event e is preceded, in the same trace, by an event resulting from executing A^{-1}.

To perform these experiments, $p_e(A|e^{-1})$ and $p_e(A|A^{-1})$ were obtained by evaluating statistics on the training set, in the same way as discussed for $p_e(A)$. The diagrams in Fig. 3(*a*) and (*b*) refer to the open and the closed world assumption, respectively, and show that the Monte Carlo approaches are much more accurate than the terms of comparison. Interestingly, they also show that the accuracy of our approach can benefit from the use of bigram models. However, the improvements of the variants MC' and MC" over MC are almost negligible under the closed world assumption: this suggests that the conditioning paradigm is effective in aligning the probabilities of the interpretations towards the ground truth, even if the starting probabilities derive from a coarser knowledge. In any case, the good accuracies achieved with the use of $p_e(A|A^{-1})$ open an interesting direction of future work: the use of Hidden Markov Models (where this form of bigram can be encoded into transitions) to support the interpretation of the traces.

Discussion. The results show that our Monte-Carlo classification algorithm is generally faster than the exhaustive approach, and that, differently from the exhaustive approach, it is feasible even over "long" traces. Furthermore, the price to be paid is negligible, as the accuracy of our algorithm is very high. Note that we considered traces even longer than the standard business process traces: in fact, in the service agency dataset traces are rarely longer than 60 (this characteristics is shared with other real datasets, such as [11]). Interestingly, we point out that the low execution times of our algorithm make our approach useful both in offline and interactive analysis. The execution times of the order of seconds back the use of our algorithm for supporting the detection of security breaches during the process-monitoring in real time: in fact, the execution of any activity in a business process typically takes more than a few seconds.

6   Related Work

Our model-driven detection of security breaches shares some connection with the activity detection and situation awareness [10] problems, and with some intrusion/threat detection approaches relying on event-based attack models (e.g., attack tree/graphs [14]. Notably, the system presented in [10] allows the analyst to monitor multiple FSMs modeling malicious/undesired behavior, by maintaining multiple "interpretation" hypotheses concurrently. However, all of the above approaches are process-unaware, in that they do not take into any account background information on process structure (expressed, e.g., via inter-activity constraints as in our work). Moreover, most of them adopt purely sequential models with no parallel branches, or just consider inter-process concurrency.

Process-awareness is a central issue instead in the areas of Business Process Management (BPM) and Business Process Intelligence (BPI). Several efforts were done in the last decade in order to integrate risk-oriented analysis mechanisms into the design, monitoring and ex-post analysis of business processes (see [18] for a recent survey on this topic). As to security-oriented risks, several log-centered approaches were proposed recently [4,5,9,13,19] that leverage different kinds of techniques developed in the Process Mining community. In particular, the usage of workflow induction [1] (resp., of compliance checking [3] and conformance checking [15]) techniques was proposed in [5] (resp., in [4]) as a way to carry out security-oriented analyses over large amounts of log data. The recognition of high-risk process instances was also faced by using a model of forbidden/undesired behavior [17], and evaluating whether a trace is an instance of it. However, to the best of our knowledge, there is no solution for the detection of security breaches that can apply to low-level logs.

Notably our idea of exploiting constraint-based process models as a form of background knowledge (especially in the "closed-world assumption" setting) is novel in the context of security-breach detection. As a matter of fact, the usage of a-priori inter-activity (precedence) constraints in a BPI setting has only been considered in the discovery of workflow models, in order to prune the search space [12] and obtain higher quality models.

Finally, the problem of mapping log events to pre-defined process activities was faced in [7,8]. In both cases, a semi-automated approach is devised, in order to help the analyst find a mapping between log events and process activities, and to eventually convert each original log trace into a sequence of the given activities. None of these approaches can hence solve the specific problem stated in this work, where a log trace is to be contrasted to multiple models (representing process behaviors and security breach patterns), in order to possibly interpret it as an instance of each of them. Rather than converting each log trace τ into a single sequence of high-level activities, our approach directly works with the raw representation of τ, and estimates the risk that τ is a security-breach instance by considering (a representative subset of) all the activity sequences that could have generated τ.

7   Conclusions and Future Work

We have proposed a probabilistic approach exploiting the knowledge of process and security-breach models for classifying business log traces as process instances and/or potential security breaches.

As mentioned in Sect. 2, the approach relies on representing the behavior of both kind of models in terms of simple constraint-oriented composition rules, which are similar in the spirit to those of declarative process modeling frameworks [2].

Clearly, such composition rule models have a limited expressive power than procedural languages (such as Petri nets, Event Process Chains, etc.), and may lose precision in the specification of highly-structured processes.

However, beside simplifying the presentation of our framework, the usage of a declarative process modeling approach is likely to be more effective to capture the behavior of flexible lowly-structured processes, and to express the partial behavioral specifications (e.g., compliance rules, misuse patterns, security-breach patterns) that typically occur in many security applications.

Anyway, the framework can be straightforwardly extended to deal with different languages/mechanisms for defining the process and security-breach models: allowing more expressive composition rules (e.g., where the tail is augmented with conditions describing the "context" that makes the rule active), or even automata or Petri nets, to specify these models simply requires to adapt function *checkModel*, that is orthogonal to the core of our technique. However, the impact of this modifications on the efficiency is worth investigating.

We also plan to consider a wider range of attributes for both events and activities (including, e.g., their respective durations and executors), and the composition rules will be extended in order to allow conditions involving these new attributes to be expressed. In particular, the possibility to use information on both the start time and the end time of the activities can help describe the behavior of a process/breach in a more precise way, by possibly expressing constraints on the way two activities may/must overlap over the time (e.g., strict precedence, temporal inclusion, etc.).

References

1. Van der Aalst, W., Weijters, T., Maruster, L.: Workflow mining: discovering process models from event logs. IEEE TKDE **16**(9), 1128–1142 (2004)
2. van der Aalst, W.M.P., Pesic, M., Schonenberg, H.: Declarative workflows: balancing between flexibility and support. Comput. Sci. - R&D **23**(2), 99–113 (2009)
3. van der Aalst, W.M.P., de Beer, H.T., van Dongen, B.F.: Process mining and verification of properties: an approach based on temporal logic. In: Meersman, R., Tari, Z. (eds.) OTM 2005. LNCS, vol. 3760, pp. 130–147. Springer, Heidelberg (2005)

4. Accorsi, R., Stocker, T.: On the exploitation of process mining for security audits: the conformance checking case. In: Proceedings of ACM SAC, pp. 1709–1716. ACM (2012)
5. Accorsi, R., Stocker, T., Müller, G.: On the exploitation of process mining for security audits: the process discovery case. In: Proceedings of ACM SAC, pp. 1462–1468. ACM (2013)
6. Agresti, A., Coull, B.A.: Approximate is better than "exact" for interval estimation of binomial proportions. Am. Stat. **52**(2), 119–126 (1998)
7. Baier, T., Mendling, J., Weske, M.: Bridging abstraction layers in process mining. Inf. Syst. **46**, 123–139 (2014)
8. Baier, T., Rogge-Solti, A., Weske, M., Mendling, J.: Matching of events and activities - an approach based on constraint satisfaction. In: Frank, U., Loucopoulos, P., Pastor, Ó., Petrounias, I. (eds.) PoEM 2014. LNBIP, vol. 197, pp. 58–72. Springer, Heidelberg (2014)
9. Bose, R., van der Aalst, W.M.: Discovering signature patterns from event logs. In: Symposium on Computational Intelligence and Data Mining (CIDM), pp. 111–118 (2013)
10. Cybenko, G., Berk, V.H.: Process query systems. IEEE Comput. **40**(1), 62–70 (2007)
11. van Dongen, B.: BPI challenge 2014: Activity log for incidents (2014). http://dx.org/10.4121/uuid:86977bac-f874-49cf-8337-80f26bf5d2ef
12. Greco, G., Guzzo, A., Lupia, F., Pontieri, L.: Process discovery under precedence constraints. ACM Trans. Knowl. Discov. Data **9**(4), 32:1–32:39 (2015)
13. Jans, M., van der Werf, J., Lybaert, N., Vanhoof, K.: A business process mining application for internal transaction fraud mitigation. Expert Syst. Appl. **38**(10), 13351–13359 (2011)
14. Lippmann, R.P., Ingols, K.W.: An annotated review of past papers on attack graphs. Technical report, DTIC Document (2005)
15. Rozinat, A., van der Aalst, W.M.: Conformance checking of processes based on monitoring real behavior. Inf. Syst. **33**(1), 64–95 (2008)
16. Sadiq, S.W., Orlowska, M.E., Sadiq, W.: Specification and validation of process constraints for flexible workflows. Inf. Syst. **30**(5), 349–378 (2005)
17. Sauer, T., Minor, M., Bergmann, R.: Inverse workflows for supporting agile business process management. In: Wissensmanagement, pp. 204–213 (2011)
18. Suriadi, S., Weiß, B., Winkelmann, A., ter Hofstede, A.H., Adams, M., Conforti, R., Fidge, C., La Rosa, M., Ouyang, C., Rosemann, M., et al.: Current research in risk-aware business process management: overview, comparison, and gap analysis. CAIS **34**(1), 933–984 (2014)
19. Werner-Stark, A., Dulai, T.: Agent-based analysis and detection of functional faults of vehicle industry processes: a process mining approach. In: Jezic, G., Kusek, M., Nguyen, N.-T., Howlett, R.J., Jain, L.C. (eds.) KES-AMSTA 2012. LNCS, vol. 7327, pp. 424–433. Springer, Heidelberg (2012)

Redescription Mining with Multi-target Predictive Clustering Trees

Matej Mihelčić[1,3]([✉]), Sašo Džeroski[2,3], Nada Lavrač[2,3], and Tomislav Šmuc[1]

[1] Ruđer Bošković Institute, Bijenička cesta 54, 10000 Zagreb, Croatia
{matej.mihelcic,tomislav.smuc}@irb.hr
[2] Jožef Stefan Institute, Jamova cesta 39, 1000 Ljubljana, Slovenia
{saso.dzeroski,nada.lavrac}@ijs.si
[3] Jožef Stefan International Postgraduate School, Jamova cesta 39,
1000 Ljubljana, Slovenia

Abstract. Redescription mining is a field of knowledge discovery that aims to find different descriptions of subsets of elements in the data by using two or more disjoint sets of descriptive attributes. The ability to find connections between different sets of descriptive attributes and provide a more comprehensive set of rules makes it very useful in practice. In this work, we introduce redescription mining algorithm for generating and iteratively improving a redescription set of user defined size based on multi-target Predictive Clustering Trees. This approach uses information about element membership in different generated rules to search for new redescriptions and is able to produce highly accurate, statistically significant redescriptions described by Boolean, nominal or numeric attributes. As opposed to current tree-based approaches that use multiclass or binary classification, we explore benefits of using multi target classification and regression to create redescriptions. The process of iterative redescription set improvement is illustrated on the dataset describing 199 world countries and their trading patterns. The performance of the algorithm is compared against the state of the art redescription mining algorithms.

Keywords: Knowledge discovery · Redescription mining · Predictive clustering trees · World countries

1 Introduction

Pattern mining [1,13] aims at discovering descriptive rules learned from data. Redescription mining [19] shares this goal but tries to find different descriptions of patterns by using two or more disjoint sets of descriptive attributes which are finally presented to the user. It is an unsupervised, descriptive knowledge discovery task. This analysis allows finding similarities between different elements and connections between different descriptive attribute sets (views) which ultimately lead to better understanding of the underlying data. Redescription mining is highly applicable in biology, economy, pharmacy, ecology and many

© Springer International Publishing Switzerland 2016
M. Ceci et al. (Eds.): NFMCP 2015, LNAI 9607, pp. 125–143, 2016.
DOI: 10.1007/978-3-319-39315-5_9

other fields, where it is important to understand connections between different descriptors and to find regularities that are valid for different element subsets. Redescriptions are represented in the form of rules and the aim is to make these rules understandable and interpretable.

The field of redescription mining was introduced by Ramakrishnan et al. [19]. Their paper presents a novel algorithm to mine redescriptions based on decision trees, called the CARTwheels. The algorithm works by building two decision trees (one for each view) that are joined in the leaves. Redescriptions are found by examining the paths from the root node of the first tree to the root node of the second and the algorithm uses multi class classification to guide the search between the two views. Other approaches to mine redescriptions include approach proposed by Zaki and Ramakrishnan [23] which uses a lattice of closed descriptor sets to find redescriptions. Further, Parida and Ramakrishnan [17] introduce algorithms for mining exact and approximate redescriptions, Gallo et al. [10] present the greedy and the MID algorithm based on frequent itemset mining.

Galbrun and Miettinen [6] present a novel greedy algorithm for mining redescriptions. In this work they extend the greedy approach by Gallo et al. [10] to work on numeric data since all previous approaches worked only on Boolean data. Redescription mining was extended by Galbrun and Kimming to a relational [5] and by Galbrun and Miettinen to the interactive setting [8]. Recently, two novel tree-based algorithms were proposed by Zinchenko [24], which explore using decision trees in a non-Boolean setting and present different methods of layer by layer tree construction, which allows making informed splits based on nodes at each level of the tree.

In this work, we explore creation and iterative improvement of redescription sets containing a user defined number of redescriptions. With this goal in mind, we developed a novel algorithm for mining redescriptions based on multi-target predictive clustering trees (PCTs) [3,14]. Our approach uses multi-target classification or regression to find highly accurate, statistically significant redescriptions, which differentiates it from other tree based approaches, especially the CARTwheels approach. Each node in a tree represents a separate rule that is used as a target in the construction of a PCT from the opposite view. Using multi-target PCTs allows us to build one model to find multiple redescriptions using nodes at all levels of the tree, further it allows to find features that are connected with multiple target features (rules) and finally due to inductive transfer [18], multi-target trees can outperform single label classification or regression trees. We have developed a procedure for rule minimization that allows us to find the smallest subset of attributes that describe a given pattern, thus we have the ability to get shorter rules even when using trees of bigger depth size. The approach is related to multi-view [2] and multilayer [11] clustering, though the main goal here is to find accurate redescriptions of interesting subsets of data, while clustering tends to find clusters that are not always easy to interpret.

After introducing the necessary notation (Sect. 2), we present the algorithm, introduce the procedure for rule minimization and perform the run-time analysis of redescription mining process (Sect. 3). We use the algorithm to iteratively improve redescription set describing 199 different world countries based on their trading behaviour [21] and general country information [22] for the year 2012 (Sect. 4). The main focus is on rules containing only logical conjunction operators, since these rules are the most interpretable and very easy to understand. In Sect. 5 we analyse redescription sets mined with one state of the art redescription mining algorithm, optimize redescription sets of equal size with our approach, compare these sets by using several criteria and discuss the results. Finally, we conclude and outline directions for future work in Sect. 6.

2   Notation and Definitions

Redescription mining in general considers redescriptions constructed on a set of views $\{W_1, W_2, \ldots, W_n\}$, however in this paper we use only two views $\{W_1, W_2\}$. The corresponding attribute (variable) sets are denoted by V_1 and V_2. Each view contains the same set of $|E|$ elements and two different sets of attributes of size $|V_1|$ and $|V_2|$. Value $W_1(i,j)$ is the value of element e_i for the attribute a_j in view W_1. The data $D = (V_1, V_2, E, W_1, W_2)$ is a quintuple of the attribute sets, the element set, and the appropriate view mappings. A query (denoted q) is a logical formula F, where q_1 contains literals from V_1. The set of elements described by a query is called its support. A redescription $R = (q_1, q_2)$ is defined as a pair of queries, one for each view in the data. The support of a redescription is the set of elements supported by both queries that constitute this redescription: $supp(R) = supp(q_1) \cap supp(q_2)$. We use $attr(R)$ to denote the multiset of attributes used in the redescription R. The accuracy of a redescription $R = (q_1, q_2)$ is measured using the Jaccard coefficient (Jaccard similarity index):

$$JS(R) = \frac{|supp(q_1) \cap supp(q_2))|}{|supp(q_1) \cup supp(q_2)|}$$

The Jaccard coefficient is not the only measure used in the field because it is possible to obtain redescriptions covering huge element subsets that necessarily have very good overlap of their queries. In this cases it is preferred to have redescriptions that reveal some more specific knowledge about the studied problem that is harder to obtain by random sampling from the underlying data distribution. This is why we compute the statistical significance (p-value) of each obtained redescription. We denote the marginal probability of a query q_1, q_2 with $p_1 = \frac{supp(q_1)}{|E|}$ and $p_2 = \frac{supp(q_2)}{|E|}$ respectively. We define the set of elements in the intersection of the queries with $o = supp(q_1) \cap supp(q_2)$. The corresponding p-value [9] is defined as

$$pV(q_1, q_2) = \sum_{n=|o|}^{|E|} \binom{|E|}{n} (p_1 \cdot p_2)^n \cdot (1 - p_1 \cdot p_2)^{|E|-n}$$

The p-value tells us if we can dismiss the null hypothesis that assumes that we obtained a given subset of elements by joining two random rules with marginal probabilities equal to the fraction of covered elements. If the obtained p-value is lower than some predefined threshold, called the significance level, then this null hypothesis should be rejected. This is a somewhat optimistic criterion, since the assumption that all elements can be sampled with equal probability need not hold for all datasets.

3   The CLUS-RM Algorithm

In this section, we describe the algorithm for mining redescriptions named CLUS-RM, that at each step improves the redescription set of the size defined by the user. The algorithm uses multi-target predictive clustering trees (PCTs) [3,14] to create a cluster hierarchy that is later transformed into redescriptions. We start by explaining the pseudo code of the algorithm (Algorithm 1) and then go into the details of each procedure in the algorithm.

Algorithm 1. The CLUS-RM algorithm

Input: First view data (W_1), Second view data (W_2), Settings file
Output: A set of redescriptions $\mathcal{R}$
 1: **procedure** CLUS-RM
 2:     $[\mathcal{D}_{W_1 init}, \mathcal{D}_{W_2 init}] \leftarrow$ prepareTargetsForInitialPCT(W_1,W_2)
 3:     [PCTW$_1$, PCTW$_2$] $\leftarrow$ createSidesInitialPCT($\mathcal{D}_{W_1 init}$, $\mathcal{D}_{W_2 init}$)
 4:     [RW$_1$, RW$_2$] $\leftarrow$ extractRules(PCTW$_1$, PCTW$_2$)
 5:     initializeArrays(elFreq, attrFreq, redScoreEl, redScoreAt, numEx, numAttr,
                    numRetRed)
 6:     **while** RunInd<maxIter **do**
 7:         [TmpRW$_1$, TmpRW$_2$] $\leftarrow$ emptyRuleSet()
 8:         [$\mathcal{D}_{W_1 Targ}$, $\mathcal{D}_{W_2 Targ}$] $\leftarrow$ prepareTargets(RW$_2$, RW$_1$)
 9:         [PCTW$_1$, PCTW$_2$] $\leftarrow$ createPCT($\mathcal{D}_{W_1 Targ}$, $\mathcal{D}_{W_2 Targ}$)
10:         TmpRW$_1$ $\leftarrow$ TmpRW$_1$ $\cup_*$ extractRules(PCTW$_1$)
11:         TmpRW$_2$ $\leftarrow$ TmpRW$_2$ $\cup_*$ extractRules(PCTW$_2$)
12:         RW$_1$ $\leftarrow$ RW$_1$ $\cup$ TmpRW$_1$
13:         RW$_2$ $\leftarrow$ RW$_2$ $\cup$ TmpRW$_2$
14:         $\mathcal{R} \leftarrow$ MineRed(RW$_1$, RW$_2$, expansionType,
                ConstSet, iteration, opSet, elFreq, attrFreq, redScoreEl, redScoreAt)
15:     **return** $\mathcal{R}$

The algorithm starts by creating initial clusters for both views (line 2 and 3 in Algorithm 1) which is achieved by transforming a non-labeled dataset into a labeled dataset of positive elements and artificially generated negative elements. For each element in the original view, we construct one negative, synthetic element (see Fig. 1) in such a way so that the original correlations among the attributes are broken. We achieve this by random shuffling of attribute values between the elements. The procedure allows experimentation with the number of shuffling steps and the number of attributes that are copied from the original elements to the artificial element. Complete randomization is achieved when the number of shuffling steps equals the number of attributes in the dataset and exactly one attribute value is copied to the artificial element at each step from a randomly chosen original element. The original elements are assigned a target label of 1.0, while the artificial elements are assigned a target label of 0.0 (see Table 1). The division between the original and the artificial elements (the idea previously used in [11]), allows us to construct a cluster hierarchy, simultaneously creating descriptions of the original elements. The described procedure is one possible way to construct the initial clusters; other approaches include assigning a random target attribute or using clusters computed by some other clustering algorithm. However, the initialization procedure used in our algorithm should preserve any strong (specific) connections and correlations that exist in the original data which are broken by using an approach that assigns random target labels.

Table 1. Creation of artificial elements for the random initialization procedure.

(a) Original dataset for view 1

Entity	W_1A_1	W_1A_2	W_1A_3
E_1	1.1	2.5	3.4
E_2	1.5	2.2	4.0
E_3	5.5	-0.6	-0.2
E_4	4.4	-0.2	2.0
E_5	3.2	1.7	2.9

(b) Original dataset for view 2

Entity	W_2A_1	W_2A_2	W_2A_3
E_1	TRUE	FALSE	FALSE
E_2	TRUE	TRUE	FALSE
E_3	FALSE	FALSE	TRUE
E_4	TRUE	TRUE	TRUE
E_5	TRUE	FALSE	TRUE

(c) Initial dataset for view 1

Entity	W_1A_1	W_1A_2	W_1A_3	Target
E_1	1.1	2.5	3.4	1.0
E_2	1.5	2.2	4.0	1.0
E_3	5.5	-0.6	-0.2	1.0
E_4	4.4	-0.2	2.0	1.0
E_5	3.2	1.7	2.9	1.0
E_1'	4.4	2.5	2.9	0.0
E_2'	3.2	-0.6	4.0	0.0
E_3'	3.2	-0.6	2.9	0.0
E_4'	4.4	-0.2	4.0	0.0
E_5'	5.5	1.7	2.9	0.0

(d) Initial dataset for view 2

Entity	W_2A_1	W_2A_2	W_2A_3	Target
E_1	TRUE	FALSE	FALSE	1.0
E_2	TRUE	TRUE	FALSE	1.0
E_3	FALSE	FALSE	TRUE	1.0
E_4	TRUE	TRUE	TRUE	1.0
E_5	TRUE	FALSE	TRUE	1.0
E_1'	TRUE	FALSE	TRUE	0.0
E_2'	FALSE	FALSE	TRUE	0.0
E_3'	TRUE	TRUE	TRUE	0.0
E_4'	FALSE	TRUE	FALSE	0.0
E_5'	FALSE	FALSE	TRUE	0.0

After creating the initial dataset, we build predictive clustering trees on both views by performing regression on the target label and using other attributes as descriptive. The decision to use regression trees instead of decision trees is purely technical, since it generates more rules because of the additional threshold associated with the target variable. These trees are converted to rules that describe element sets and are necessary for the next step of the algorithm. The rule lists RW_1 and RW_2 contain generated rules, and a new rule is added to the list if it differs from all other rules in a predefined number of attributes or if it describes a new unique element subset (the $\cup_*$ operator in Algorithm 1). The iterative process of the algorithm begins right after rule creation. Here, we create targets based on the rules obtained in the previous step or in the initialization step. The Rules obtained by predictive clustering on W_1 are used to build targets for clustering on W_2 (denoted W_1T_1, W_1T_2), and vice versa. For each element in the dataset we assign label 1.0 if the element is described by some specific rule, otherwise 0.0 (see Table 2). For example, the attribute W_2T_1 from dataset for view 1 represents the condition $IF\ W_2A_1 = TRUE$ (constructed on dataset for view 2), which describes elements E_1, E_2, E_4, E_5. By placing this target attribute in the view 1 dataset, we guide the PCT construction to create a cluster containing and describing the same set of elements with descriptive variables of view 1 (a choice that satisfies this condition is $IF\ W_1A_3 > 0$).

Table 2. Intermediate generation of labels based on discovered rules.

(a) Dataset for view 1

E	W_1A_1	W_1A_2	W_1A_3	W_2T_1	W_2T_2
E_1	1.1	2.5	3.4	1.0	0.0
E_2	1.5	2.2	4.0	1.0	0.0
E_3	5.5	-0.6	-0.2	0.0	0.0
E_4	4.4	-0.2	2.0	1.0	0.0
E_5	3.2	1.7	2.9	1.0	1.0

(b) Dataset for view 2

E	W_2A_1	W_2A_2	W_2A_3	W_1T_1	W_1T_2
E_1	TRUE	FALSE	FALSE	0.0	1.0
E_2	TRUE	TRUE	FALSE	0.0	1.0
E_3	FALSE	FALSE	TRUE	1.0	0.0
E_4	TRUE	TRUE	TRUE	1.0	0.0
E_5	TRUE	FALSE	TRUE	1.0	1.0

Rules obtained in the previous step are combined into redescriptions if they satisfy a given set of constraints $ConstSet$. The set of constraints consists of minimal Jaccard coefficient ($minJS$), maximum allowed p-value ($maxPval$) and minimum and maximum support ($minSupp$, $maxSupp$) which have to be satisfied for a redescription to be considered as a candidate for the redescription set.

3.1   The Procedure for Creating Redescriptions

The algorithm for creating redescriptions from rules (Algorithm 2) joins view 1 rules (or its negation, if allowed by the user) with rules (or its negation) from view 2 (see Fig. 1 and line 2 in Algorithm 2). We distinguish three cases of creating redescriptions from rules (expansion types):

1. Unguided initial: $UInit \leftarrow (RW_1 \times_{ConstSet}^{opSet\backslash\{\vee\}} RW_2)$
2. Unguided: $U \leftarrow (RW_{1_{newRuleIt}} \times_{ConstSet}^{opSet\backslash\{\vee\}} RW_{2_{newRuleIt}})$
3. Guided: $G \leftarrow (RW_{1_{newRuleIt}} \times_{ConstSet}^{opSet\backslash\{\vee\}} RW_{2_{oldRuleIt}}) \cup$
   $(RW_{1_{oldRuleIt}} \times_{ConstSet}^{opSet\backslash\{\vee\}} RW_{2_{newRuleIt}})$

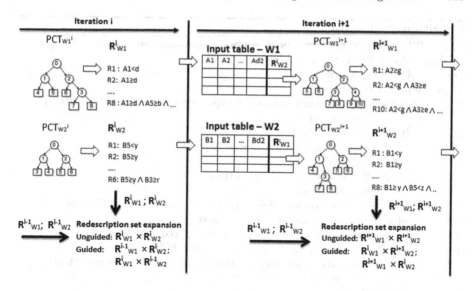

Fig. 1. Illustration of rule, redescription construction and iterations

The $\times_{ConstSet}^{opSet}$ operator denotes a Cartesian product of two sets, allowing the use of logical operators from *opSet* and leaving only those redescriptions that satisfy a given set of constraints *ConstSet*. The unguided expansion allows obtaining redescriptions with more diverse subsets of elements that can later be improved through the iteration process.

The algorithm finds first *numRed* redescriptions and then iteratively enriches this set by exchanging the redescription with the worst comparative score with the newly created redescription (lines 3-14 in Algorithm 2). The algorithm uses 4 arrays (*elFreq, attrFreq, redScoreEl, redScoreAt*) to incrementally add and improve redescriptions in the redescription set. The element/attribute frequency arrays contain the number of times each element/attribute from the dataset occurs in redescriptions from a redescription set. Redescription scores are computed as $redScoreEl(R) = \sum_{e \in supp(R)} (elFreq[e] - 1)$, and $redScoreAt(R) = \sum_{a \in attr(R)} (attrFreq[a] - 1)$. The score of a new redescription is computed in the same way by using existing frequencies from the set. If the algorithm finds a redescription R' such that $R_i = argmax_{R \in \mathcal{R}|\ R.pval \geq R'.pval} score(R', R)$, where $score(R', R) = (\frac{(1.0 - R'.elSc + 1.0 - R'.atrSc + R'.JS)}{3} - \frac{(1.0 - R.elSc + 1.0 - R.attrSc + R.JS)}{3})$, all arrays are updated so that the frequencies of elements described by R_i and attributes contained in it's queries are decreased by one, while the frequencies of elements and attributes associated with R' are increased. This score favours redescriptions that describe elements with low frequency by using non frequent attributes. At the same time it finds as accurate and significant redescriptions as possible.

Element weighting has been used before in subgroup discovery [12,15] to model covering importance for elements. Our approach is similar but uses

Algorithm 2. MineRed

Input: RW_1, RW_2, expansion type, ConstSet, iteration number, opSet, elFreq, attr-
Freq, redScoreEl, redScoreAt
Output: A set of redescriptions $\mathcal{R}$

```
 1: procedure MINERED
 2:     expansionSet ← returnExpansionSet(expansionType, opSet, RW₁, RW₂)
 3:     for R′ ∈ expansionSet do
 4:         if |R|<ConstSet.MaxRed then
 5:             updateFrequencies(elFreq, attrFreq, R′)
 6:             R ← R ∪ R′
 7:             if |R| == ConstSet.MaxRed then
 8:                 for R ∈ R do
 9:                     computeScores(elFreq,attrFreq, redScoreEl, redScoreAt, R)
10:         else if |R| == ConstSet.MaxRed then
11:             compScore(elFreq,attrFreq, redScoreEl, redScoreAt, R′)
12:             Rb ← argmaxR∈R| R.pval≥R′.pval score(R′, R)
13:             updtFreqAndScores(elFreq, attrFreq, redScoreEl, redScoreAt, R′, R)
14:             R ← R\Rb ∪ R′
15:     if ∨ ∈ opSet then
16:         for R ∈ R do
17:             if expansionType==unguidedExpansion AND iteration==0 then
18:                 ind ← 0
19:             else
20:                 ind ← newRuleIt
21:             r′W₁ ← argmax(R.maxRef(r), R.maxRef(¬r), r ∈ RW1ind)
22:             Rref ← (r′W₁ ∨ R.rW₁ × R.rW₂)
23:             r′W₂ ← argmax(Rref.maxRef(r), Rref.maxRef(¬r), r ∈ RW2ind)
24:             Rref ← (Rref.rW₁ × r′W₂ ∨ R.rW₂)
25:             updtFreqAndScores(elFreq, attrFreq, redScoreEl, redScoreAt, R, Rref)
26:             R ← R\R ∪ Rref
27:     return R
```

---

different weighting mechanism, adapts it to the redescription mining setting by combining element and attribute weights and incorporates it into the framework of iterative redescription set refinement in which some redescriptions can be replaced with more suitable candidates.

The algorithm can use three types of logical operators (disjunction, conjunction and negation). The disjunction operator is used to increase redescription accuracy and support (lines 15–26 in Algorithm 2). For a redescription $R = (q_1, q_2)$, we find rules r that maximize:

1. $JS(supp(q_1 \vee r)\backslash supp(R), supp(q_2)\backslash supp(R))$
2. $JS(supp(q_1 \vee \neg r)\backslash supp(R), supp(q_2)\backslash supp(R))$
3. $JS(supp(q_1)\backslash supp(R), supp(q_2 \vee r)\backslash supp(R))$
4. $JS(supp(q_1)\backslash supp(R), supp(q_2 \vee \neg r)\backslash supp(R))$

The rule r is found so that it covers elements that are supported by q_2 but not by q_1 ($R.maxRef(r')$, $r' \in RW_1$) and vice versa.

3.2   Rule Size Minimization

Rule minimization procedure is applied in the final step of redescription set creation. The main goal of this procedure is to find a minimal attribute set for all rules contained in redescriptions that describe the same pattern as the original redescription. This leads to better understandability and readability of returned redescriptions.

The method minimizes conjunctive formulas $F = v_1 \wedge v_2 \wedge v_3 \wedge v_4 \wedge \cdots \wedge v_n$, where each v_i denotes one literal of the form $v_i = c$ in the case of Boolean or categorical attributes or $c_1 \leq v_i \leq c_2$ in the case of numerical attributes. The procedure chooses each v_i in turn, computes $\mathcal{S}_{v_i} = supp(v_i) \backslash supp(F)$ and then finds the minimal set $\mathcal{T} = \{v_k, \ldots, v_m\}$ such that $\forall e \in \mathcal{S}_{v_i}, \exists v_j \in \mathcal{T}, e \notin supp(v_j)$ and $\cap_k v_k = supp(F)$, $v_k \in \mathcal{T}$ (see Fig. 2). The procedure returns a family of sets $\mathcal{F} = \{\mathcal{T}_i, i = 1, \ldots, n\}$ and chooses the representative set containing the smallest number of attributes.

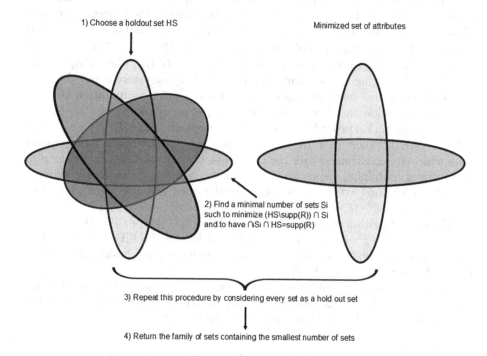

1) Choose a holdout set HS

Minimized set of attributes

2) Find a minimal number of sets Si such to minimize (HS\supp(R)) ∩ Si and to have ∩Si ∩ HS=supp(R)

3) Repeat this procedure by considering every set as a hold out set

4) Return the family of sets containing the smallest number of sets

Fig. 2. Rule minimization procedure

The procedure is related to a procedure for finding a minimal set of generators in [23]. It is constructed with a purpose of minimizing rules contained in already constructed redescriptions whereas minimal set of generators is used to construct redescriptions which requires it to compute a closed lattice of descriptors.

3.3 Algorithm Time Complexity

In this subsection we analyse the algorithm's time complexity. We start from the known results [20] that predictive clustering tree construction has the worst time complexity of $O(z \cdot m \cdot |E|^2)$ to completely induce the tree, where m denotes the number of descriptive variables in a selected view and z the total number of internal nodes in the tree.

We use the HashSet and the HashMap data structure with open addressing to store elements which have the time complexity of $O(1)$ for add, remove, contains and size assuming the hash function behaves in a random enough manner (uniform hashing).

The initialization step has the complexity of $O(|E| \cdot (|V_1| + |V_2|))$ and the PCT to rules transformation has the complexity of $O(z)$. Creation of redescriptions via extraction/filtering of pairs obtained from Cartesian product of two rule sets has the worst time complexity of $O(n + n')$, where n equals the number of elements covered by the rule created on $W1$ and n' denotes the number of elements covered by the rule created on $W2$. To compute the Cartesian product of two rule sets we make $\sum_{i \in R_L} \sum_{j \in R_R} (n_i + n_j)$ steps. As both $n \leq |E|$ and $n' \leq |E|$, the worst time complexity of this step is $O(z^2 \cdot |E|)$. However, if we have a balanced tree, the complexity is closer to $O(z \cdot d \cdot |E|)$, where d equals the tree depth. Updating the attribute and element frequency tables and the total redescription scores has the complexity of $O(|E|)$. The computation of rules containing negation and disjunction operators has a complexity of $O(z \cdot |E|)$.

The minimization procedure has the time complexity of $O(|\mathcal{R}| \cdot ((a + a') \cdot |E| + (a^3 + a'^3) \cdot |E|))$, where a, a' represent the number of attributes in redescription rules which are constrained with the tree depth d (or a constant multiple of d in case of rules containing disjunctions). As the the number of elements in support of such constrained attributes is much smaller then $|E|$, the worst case time complexity is $O(d^3 \cdot |E|)$.

The total algorithm time complexity is: $O(|E| \cdot (|V_1| + |V_2|) + z \cdot |V_1| \cdot |E|^2 + z \cdot |V_2| \cdot |E|^2 + 2 \cdot z + z^2 \cdot |E| + z^2 \cdot |E| + 2 \cdot z \cdot |E| + d^3 \cdot |E|)$ which is $O(z \cdot (|V_1| + |V_2|) \cdot |E|^2 + z^2 \cdot |E|)$. The pessimistic worst time complexity assuming inadequate hashing function is $O(z^2 \cdot |E|^2 + z \cdot (|V_1| + |V_2|) \cdot |E|^2)$.

Optimizations that could speed up computing redescriptions include the use of rule indexing that would allow combining only those rules certain to cross the user defined thresholds and Local Sensitive Hashing [4].

4   Mining Redescriptions on Data Describing Countries

We present the experimental results of mining redescriptions with our algorithm on data describing 199 world countries in the year 2012 ([11, 21, 22]). The dataset has two views, both containing numerical attributes with possible missing values. One view contains 312 attributes representing the importance of import and export of different commodities for countries, while the second view contains 49 attributes with country information provided by the World Bank.

There are several techniques described in [6] for computing Jaccard coefficient when data contains missing values. We compute the Jaccard coefficient guided by the principle that an element can not be in a support of a rule containing only conjunction operator if it has missing values for some of the attributes contained in a condition of a rule. We use notation from [6] to denote $E_{1,1} = supp(q_1) \cap supp(q_2)$, $E_{1,0} = supp(q_1) \setminus supp(q_2)$, $E_{0,1} = supp(q_2) \setminus supp(q_1)$, $E_{1,?} = supp(q_1) \cap missing(q_2)$, $E_{?,1} = missing(q_1) \cap supp(q_2)$, where $R = (q_1, q_2)$ and $missing(q)$ represents a set of elements for which we can not determine if they are in support of q due to missing values. We define the Jaccard coefficient as:

$$JS_m(R) = \frac{|E_{1,1}|}{|E_{1,0}| + |E_{0,1}| + |E_{1,1}| + |E_{1,?}| + |E_{?,1}|}$$

It holds that $JS_{pes}(R) \le JS_m(R) \le JS_{opt}(R)$, where JS_{opt} and JS_{pes} denote optimistic and pessimistic estimate of JS when dealing with missing values.

The algorithm was tested with 50, 200, 800 iterations and only rules containing the conjunction operator were allowed. For each number of iterations, we performed 10 runs of the algorithm, computed redescription sets containing 50 redescriptions and measured the average Jaccard coefficient and the average redescription support. Allowed redescription supports were in range $[5, 120]$, the maximum p-value equalled 0.01 and the minimum Jaccard coefficient was 0.6. We used complete randomization in initialization procedure.

Figure 3 shows that with increased number of iterations, the algorithm finds redescriptions with higher accuracy, but describing smaller subsets of countries. The mean value of the total overall coverage of elements in the redescription set varies between 47 % and 53 %. This indicates that the algorithm managed to find highly accurate redescriptions describing a significant number of total elements from the dataset.

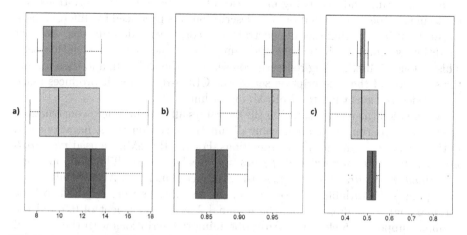

Fig. 3. A summary of the results for different numbers of algorithm runs (top to bottom: 800, 200, 50): average redescription support size a), average Jaccard coefficient b), fraction of all elements described by a redescription c).

We demonstrate one highly accurate, statistically significant redescription mined on the Country dataset. Several additional examples can be seen in [16].

```
W1R: EMPL_BAD >= 5.6 <= 12.5 AND POP_14 >= 13.166 <= 18.2591 AND
     CRED >= 99.2251 <= 305.0869
W2R: E/I_MED_PH >= 0.927 <= 4.563 AND E/I_FB >= 0.381 <= 1.46 AND
     E/I_PULP_WP >= 0.332 <= 859.221
```

This redescription describes 14 world countries (United Kingdom, Switzerland, Sweden, Spain, Singapore, Netherlands, Malta, Luxembourg, Germany, France, Finland, Denmark, Cyprus and Austria) with Jaccard coefficient 1.0. We found that the vulnerable employment of these countries ranges from $[5.6, 12.5]\%$, the percentage of population aged $0 - 14$ is in $[13.2, 18.3]\%$, and the domestic credit to private sector is in $[99.2, 305.1]\%$ of the GDP. In addition, export to import ratio of medicinal and pharmaceutical products is in $[0.9, 4.6]\%$, export to import ratio of basic food is in $[0.4, 1.5]\%$ and this ratio for pulp and waste paper is in $[0.3, 859.2]\%$. This is a statistically highly significant redescription with a p-value of $1.5 \cdot 10^{-13}$, it contains 3 descriptive variables for view 1 and 3 variables for view 2. It is a medium size redescription, based on its rule size.

5   Algorithm Evaluation and Comparison

In this section, we compare rules produced by our algorithm with the current state of the art algorithm ReReMi, described in [9]. We used the Siren tool [7] to perform redescription mining with the ReReMi algorithm on the Country dataset described in Sect. 4. The layered/split tree algorithms (described in [24]) currently do not work with data that contain missing values.

Redescription mining algorithm comparison was mainly done in the literature by selecting and discussing properties of the individual redescriptions. We try to make objective evaluation of redescription sets produced by different algorithms by using the same set of constraints to construct redescriptions. Another condition we imposed is to have the same size of the final redescription sets. This is done by first finding redescription set with the ReReMi, and then forcing the same size of the redescription set on the CLUS-RM, since it produces much more redescriptions than the ReReMi algorithm.

We divided the results based on the operators allowed for query construction. In the first experiment we allow using disjunctions, conjunctions, negations and in the second experiment only conjunctions. For the ReReMi, we used *max product buckets = 200*, *max number of pairs = 500* when using all logical operators, *max number of pairs = 1000* when using only conjunctions. Also, we allowed a maximum of 15 variables for each query. Redescriptions were required to have the maximal $p - value$ of 0.01, the minimal Jaccard coefficient of 0.5 and the minimal support of 5 elements. After obtaining redescriptions with the ReReMi algorithm, we used the *Filter redundant redescriptions* option to remove duplicate and redundant redescriptions with the *max overlap option* equal to 0.99. For each redescription set, we optimized a redescription set of the same size by using the CLUS-RM algorithm. We used 800 iterations keeping constraints for the Jaccard coefficient, the $p-value$ and support. Maximum allowed average tree depth was set to 8 and we used the complete randomization in the initialization procedure.

For the generated redescription sets, we plot comparative boxplots for the Jaccard coefficient, the log_{10} of the $p-value$, the element overlap, the attribute overlap and the rule size. The element overlap is the average Jaccard coefficient of covered elements by one redescription with respect to all other redescriptions in the redescription set, similarly the attribute overlap is the average Jaccard coefficient of the attributes contained in the redescription queries compared to every other redescription in the set. To emphasize importance of the redescription size from the point of understandability ($|attr(R)| \geq 20$ considered to be highly complex to understand), we calculate the normalized redescription size as follows:

$$R_{size} = \begin{cases} \frac{|attr(R)|}{20} & , |attr(R)| < 20 \\ 1 & , 20 \leq |attr(R)| \end{cases}$$

To obtain comparative results, we optimized JS_{pes} with ReReMi algorithm and then recalculated the score for each redescription to obtain JS_m.

The Fig. 4 shows that the CLUS-RM found statistically significant redescriptions with Jaccard coefficient higher than those produced by the ReReMi algorithm. Due to its goal of finding highly accurate but minimally overlapping redescriptions in terms of elements and attributes, it found redescriptions with smaller support when conjunctions, disjunctions and negations are allowed. One important thing is that this was achieved by using redescriptions that mostly have smaller query size than the ReReMi produced redescriptions. We report two more statistics, the element coverage (the percentage of total elements described by at least one redescription) and attribute coverage (the percentage of attributes used in redescription rules). The CLUS-RM described 99 % of elements while ReReMi described 100 % elements. The CLUS-RM used 47 % of all attributes in the rules and the ReReMi used 41 %.

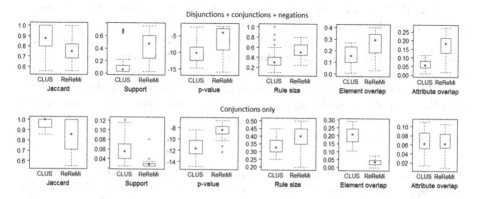

Fig. 4. The CLUS-RM and the ReReMi algorithm comparison on two redescription sets: constructed by using disjunctions, conjunctions, negations (120 redescriptions) and by using only conjunction operator (36 redescriptions)

The evaluation on the redescription sets constructed by only using conjunction operator showed that the CLUS-RM produced redescriptions with higher Jaccard coefficient, higher support and smaller $p - value$ than the ReReMi algorithm. As a consequence, the CLUS-RM has higher element overlap but also somewhat smaller query size in redescriptions. Attribute overlaps are comparable between approaches. The CLUS-RM covered 25 % elements and the ReReMi algorithm 53 % while the attribute coverage is 27 % and 36 %.

The CLUS-RM approach produces redescriptions containing mainly conjunction operators while the ReReMi approach uses mostly disjunctions if allowed. The redescription sets obtained with the CLUS-RM contained highly accurate, statistically significant, mostly non overlapping redescriptions. There are two possible techniques available to obtain redescriptions with higher support with the CLUS-RM algorithm: to increase the minimal support or to increase the redescription set size. We believe that the proposed approach complements the ReReMi approach by finding many significant conjunction based redescriptions.

6   Conclusion

This work introduces a novel redescription mining framework which optimizes a redescription set of user defined size. The algorithm is based on multi-target predictive clustering trees, which allows using element coverage by rules constructed on one view as targets for the other view. Produced redescriptions incrementally improve the redescription set by using a predefined set of criteria (the Jaccard coefficient, the p-value, the element overlap and the attribute overlap). The ability to construct many different redescriptions and use them to optimize a set of fixed size differentiates the approach from currently proposed solutions. We analysed the algorithm time complexity and measured its performance on data describing world countries. The results show that, when finding redescriptions containing only conjunction operator, there are benefits of using more iterations. Generated redescriptions are statistically relevant with p-values less than 10^{-5}. Many generated rules contained the maximum of 6 attributes per rule in a redescription. Finally, we compare some characteristics of redescription sets generated by the CLUS-RM and the ReReMi algorithms. These results and comparison reveal the main difference in algorithm preference - CLUS-RM producing more accurate redescriptions using much more conjunctive rules.

In future work, we plan to extend the current framework by deploying Random Forest of PCTs, which should further boost resulting redescription sets in terms of size, diversity and quality. We also intend to work on more comprehensive and objective evaluation of redescription sets.

Acknowledgement. The authors would like to acknowledge the European Commission's support through the MAESTRA project (Gr. no. 612944), the MULTIPLEX project (Gr.no. 317532), the InnoMol project (Gr. no. 316289), and support of the Croatian Science Foundation (Pr. no. 9623: Machine Learning Algorithms for Insightful Analysis of Complex Data Structures).

A   Appendix

We present several shorter redescriptions mined by the CLUS-RM and the
ReReMi algorithm. The full names of the attributes used in redescription queries
can be seen in Fig. 5.

Table 3. Redescription examples produced by CLUS-RM and ReReMi algorithm using
only conjunction operator

Redescriptions	JS	supp	p-value	Algorithm
$-0.04 \leq POP\_GROWTH \leq 2.49 \land$ $14.05 \leq POP\_64 \leq 21.1 \land$ $13.74 \leq RUR\_POP \leq 50.1 \land$ $4.7 \leq EMP\_PART\_M \leq 11.7$ $1.0 \leq E\_PL\_PF \leq 2.0 \land$ $0.65 \leq E/I\_SPEC\_MACH \leq 4.31 \land$ $18.0 \leq I\_MED\_S\_TIM \leq 25.0$	1.0	14	$1.5 \cdot 10^{-13}$	CLUS-RM
$26.0 < RUR\_POP \land$ $31.0 < CRED\_COVER < 47.5 \land$ $52.9 < LABOR\_F < 67.5$ $4.0 < E\_MAN\_G < 17.0 \land$ $17.0 < I\_MED\_S\_TIM < 33.0 \land$ $I\_TF\_W < 0.0 \land I\_AN\_VEG\_OIL < 1.0 \land$ $I\_TY\_RP < 2.0 \land$ $0.01 < E/I\_MED\_PH < 0.24$	1.0	8	$6.4 \cdot 10^{-11}$	ReReMi
$14.56 \leq POP\_14 \leq 21.54 \land$ $2.2 \leq MORT \leq 4.5 \land$ $61.8 \leq LABOR\_M \leq 68.1$ $1.0 \leq E/I\_OTH\_MACH\_PART \leq 1.62 \land$ $14.0 \leq I\_HIGH\_S\_TIM \leq 28.0 \land$ $0.74 \leq E/I\_PULP\_WP \leq 13.04$	1.0	9	$1.9 \cdot 10^{-11}$	CLUS-RM
$14.57 < POP\_14 < 14.98 \land$ $POP\_GROWTH < 0.26 \land$ $24.0 < UNEMPL\_YOUTH\_F < 44.3$ $9.0 < E\_FB < 14.0 \land$ $1.0 < E\_NFM \land I\_T\_TM < 0.0 \land$ $2.17 < E/I\_PEARLS\_PSM < 885.93 \land$ $0.83 < E/I\_PRIM\_COM < 2.93$	1.0	5	$4.6 \cdot 10^{-9}$	ReReMi

Table 4. Redescription examples produced by CLUS-RM and ReReMi algorithm using conjunction, disjunction and negation operators

Redescriptions	JS	supp	p-value	Algorithm
$8.0 \leq MORT \leq 181.6$ $\neg\,(0.65 \leq E/I\_SPEC\_MACH \leq 61.94)$	0.842	139	$2.6 \cdot 10^{-4}$	CLUS-RM
$4.9 < MORT \;\vee\; 22.8 < POP\_14 \;\vee$ $-0.26 < POP\_GROWTH < -0.09 \;\vee$ $11.1 < UNEMPL\_LONG$ $E\_PH\_OPT\_WT < 0.0 \;\wedge$ $E/I\_SPEC\_MACH < 0.65$	0.865	148	$6.3 \cdot 10^{-4}$	ReReMi
$-1.48 \leq POP\_GROWTH \leq 0.48 \;\wedge$ $2.9 \leq MORT \leq 5.4 \;\wedge$ $-1.49 \leq BAL \leq 10.11 \;\wedge$ $4.24 \leq STOCKS \leq 166.61$ $1.0 \leq I\_NM\_MIN\_MAN \leq 1.0 \;\wedge$ $0.81 \leq E/I\_OTH\_MACH\_PART \leq 3.15 \;\wedge$ $19.0 \leq I\_MACH\_TRANS\_EQ \leq 32.0 \;\wedge$ $0.92 \leq E/I\_CHEM\_PROD \leq 6.66$	1.0	12	$8.56 \cdot 10^{-13}$	CLUS-RM
$MORT < 95.5 \;\wedge$ $75.9 < LABOR\_F \;\wedge$ $1.89 < POP\_GROWTH$ $((94.0 < I\_ALL\_AP < 99.0 \;\wedge$ $E/I\_PL\_PF < 0.03) \;\vee$ $31.0 < I\_OR\_MET\_PS\_NMG < 34.0 \;\vee$ $5.38 < E/I\_CM\_IEF < 7.81) \;\wedge$ $1.0 < E/I\_COFF\_TEA\_SPICE$	1.0	10	$6.3 \cdot 10^{-12}$	ReReMi

In Table 3 we show two very accurate redescriptions mined with the ReReMi algorithm and compare it to two redescriptions mined with the CLUS-RM.

In Table 4, we present two redescriptions containing conjunction and disjunction operator obtained with the ReReMi algorithm, and two redescriptions containing conjunctions and negations obtained with the CLUS-RM algorithm. This examples demonstrate the main difference between the methodologies. The ReReMi algorithm uses disjunction operator often in redescription construction whereas the CLUS-RM mostly uses conjunction operator to construct redescriptions.

TRADE:

FB - Food, basic
NFM - Non-ferrous metals
PEARLS_PSM - Pearls, precious stones and non-monetary gold
CHEM_PROD - Chemical products
MACH_TRANS_EQ - Machinery and transport equipment
PRIM_COM - Primary commodities, precious stones and non-monetary gold,
          excluding fuels
OR_MET_PS_NMG - Ores, metals, precious stones and non-monetary gold
MED_S_TIM - Medium-skill and technology-intensive manufactures
HIGH_S_TIM|High-skill and technology-intensive manufactures
COFF_TEA_SPICE - Coffee, tea, cocoa, spices, and manufactures thereof
T_TM - Tobacco and tobacco manufactures
CM_IEF - Crude materials, inedible, except fuels
PULP_WP - Pulp and waste paper
TF_W - Textiles fibres and their wastes
AN_VEG_OIL - Animal and vegetable oils, fats and waxes
MED_PH - Medicinal and pharmaceutical products
PL_PF - Plastics in primary forms
MAN_G - Manufactured goods
TY_RP - Textile yarn and related products
NM_MIN_MAN - Non metallic mineral manufactures, n.e.s.
SPEC_MACH - Specialised machinery
OTH_MACH_PART - Other industrial machinery and parts
ALL_AP - All allocated products
PH_OPT_WT - Photo apparatus, optical goods, watches and clocks

COUNTRY INFORMATION:

POP_GROWTH - Population growth (annual %)
POP_64 - Population ages 65 and above (% of total)
POP_14 - Population ages 0-14 (% of total)
RUR_POP - Rural population (% of total population)
EMP_PART_M - Part time employment, male (% of total male employment)
CRED_COVER - Private credit bureau coverage (% of adults)
LABOR_F - Labor participation rate,(% female population, 15+)
LABOR_M - Labor participation rate, (% male population, 15+)
UNEMPL_YOUTH_F - Unemployment, youth (% female labor force 15-24)
MORT - Mortality rate, under-5 (per 1,000)
UNEMPL_LONG - Long-term unemployment (% of total unemployment)

Fig. 5. Indicator full names

References

1. Agrawal, R., Imieliński, T., Swami, A.: Mining association rules between sets of items in large databases. In: Proceedings of the 1993 ACM SIGMOD International Conference on Management of Data, pp. 207–216, Washington, D.C. (1993)
2. Bickel, S., Scheffer, T.: Multi-view clustering. In: Proceedings of the Fourth IEEE International Conference on Data Mining, pp. 19–26, Washington, D.C. (2004)
3. Blockeel., H.: Top-down induction of first order logical decision trees. Ph.d. thesis, Katholieke Universiteit Leuven, Department of Computer Science (1998)
4. Cohen, E., Datar, M., Fujiwara, S., Gionis, A., Indyk, P., Motwani, R., Ullman, J., D., Yang, C.: Finding interesting associations without support pruning. In: ICDE, pp. 489–499 (2000)
5. Galbrun, E., Kimmig, A.: Finding relational redescriptions. Mach. Learn. **96**, 225–248 (2014)
6. Galbrun, E., Miettinen, P.: From black and white to full color: extending redescription mining outside the Boolean world. Stat. Anal. Data Mining **5**, 284–303 (2012)
7. Galbrun, E., Miettinen, P.: Siren : An interactive tool for mining and visualizing geospatial redescriptions. In: KDD, pp. 1544–1547 (2012)
8. Galbrun, E., Miettinen, P.: A case of visual and interactive data analysis: geospatial redescription mining. In: Instant Interactive Data Mining Workshop @ ECML-PKDD (2012)
9. Galbrun, E.: Methods for redescription mining. Ph.d. thesis, University of Helsinki (2013)
10. Gallo, A., Miettinen, P., Mannila, H.: Finding subgroups having several descriptions: algorithms for redescription mining. In: Proceedings of the SIAM International Conference on Data Mining, Atlanta, Georgia, pp. 334–345 (2008)
11. Gamberger, D., Mihelčić, M., Lavrač, N.: Multilayer clustering: a discovery experiment on country level trading data. In: Džeroski, S., Panov, P., Kocev, D., Todorovski, L. (eds.) DS 2014. LNCS, vol. 8777, pp. 87–98. Springer, Heidelberg (2014)
12. Gamberger, D., Lavrač, N.: Expert-guided subgroup discovery: methodology and application. J. Artif. Intell. Res. **17**, 501–527 (2002)
13. Giacometti, A., Li, D.H., Marcel, P., Soulet, A.: 20 years of pattern mining: a bibliometric survey. SIGKDD Explor. Newsl. **15**, 41–50 (2014)
14. Kocev, D., Vens, C., Struyf, J., Džeroski, S.: Tree ensembles for predicting structured outputs. Pattern Recogn. **46**, 817–833 (2013)
15. Lavrač, N., Kavšek, B., Flach, P., Lj, T.: Subgroup discovery with CN2-SD. J. Mach. Learn. Res. **5**, 153–188 (2004)
16. Mihelčić, M., Džeroski, S., Lavrač, N., Šmuc, T.: Redescription mining with multi-label predictive clustering trees. In: Proceedings of the Fourth Workshop on New Frontiers in Mining Complex Patterns @ ECML-PKDD, pp. 86–97. Porto (2015)
17. Parida, L., Ramakrishnan, N.: Redescription mining: structure theory and algorithms. In: Proceedings of the 20th National Conference on Artificial Intelligence, Pittsburgh, Pennsylvania, pp. 837–844 (2004)
18. Piccart, B.: Algorithms for multi-target learning. Ph.d. thesis, Katholieke Universiteit Leuven (2012)
19. Ramakrishnan, N., Kumar, D., Mishra, B., Potts, M., Helm, R. F.: Turning CART-wheels: an alternating algorithm for mining redescriptions. In: Proceedings of the Tenth ACM SIGKDD International Conference on Knowledge Discovery and Data Mining, pp. 266–275. ACM, Seattle, WA (2004)

20. Stojanova, D., Ceci, M., Appice, A., Džeroski, S.: Network regression with predictive clustering trees. Data Min. Knowl. Disc. **25**, 378–413 (2012)
21. UNCTAD database. http://unctadstat.unctad.org/EN/
22. World Bank database. http://data.worldbank.org/
23. Zaki, M. J., Ramakrishnan, N.: Reasoning about sets using redescription mining. In: Proceedings of the Eleventh ACM SIGKDD International Conference on Knowledge Discovery and Data Mining, pp. 364–373. ACM, Chicago, Illinois (2005)
24. Zinchenko, T., Redescription mining over non-binary data sets using decision trees. Masters thesis, Universität des Saarlandes (2014)

Mining Complex Data

Generalizing Patterns for Cross-Domain Analogy

Fabio Leuzzi[1] and Stefano Ferilli[1,2(✉)]

[1] Dipartimento di Informatica, Università di Bari, Bari, Italy
{fabio.leuzzi,stefano.ferilli}@uniba.it
[2] Centro Interdipartimentale per la Logica e sue Applicazioni,
Università di Bari, Bari, Italy

Abstract. Analogy is the cognitive process of matching the characterizing features of two different items. This may enable reuse of knowledge across domains, which can be helpful to solve problems. Analogy is strongly related to semantics, because the mappings are based on the role and meaning of the features, which goes beyond simple syntactic association. The analogical mappings found between pairs of descriptions can be used to obtain more general analogical patterns. Such patterns may be stored in the long term memory, allowing self-improvement and growth. This paper proposes generalizations of patterns obtained by analogy, carried out through two main steps: (1) isolating analogous roles of two descriptions coming from different domains, and (2) abstracting from portions of knowledge that have no analogical relationships. The result is a multi-strategy approach in which the analogy brings to a generalization, that is, in turn, a novel description to reason over and over again. An example is provided to show the behavior and effect of the proposed generalization approach.

1 Introduction

Analogy is the cognitive process of matching the characterizing features of two different items (subjects, objects, situations, etc.). While it is often confused with similarity, there is a significant difference between them. Similarity maps exactly the same features in the two items. In analogy, a feature in one item can be mapped onto a completely different feature in the other, provided that they both play in some sense 'the same role' in the respective items. So, analogy is much related to *abstraction*, while similarity is more related to generalization. Indeed, abstracting the 'role' of the features away from their specific embodiment in the single items is fundamental to recognize the possibility of an analogical mapping between them. In some sense, similarity is a (simpler) kind of analogy. There is also some connection between analogy and metaphors, in that the latter are one-way mappings: the features of one item can be mapped onto those of the other, but the reverse mapping would not make sense. It is clear that analogy has a tight relationship to semantics, because the mappings are based on the role and meaning of the features in the two items, which goes beyond simple syntactic association (as for similarity). It allows one to reuse knowledge from a known domain to an unknown one, without having to learn

M. Ceci et al. (Eds.): NFMCP 2015, LNAI 9607, pp. 147–162, 2016.
DOI: 10.1007/978-3-319-39315-5_10

from scratch. In fact, after finding the analogy on some (fundamental) features, the association can be extended to further features, for which experience has not yet discovered the association of roles.

In everyday life people often face problems about which they have no experience. Sometimes they call a friend for help, or call an expert that is able to solve the problem. However, sometimes they are able to identify an analogous experience (or any other type of reliable knowledge), from which inferring an hypothetical solution. The inferred solution is not always applicable (e.g., it might seem unreasonable), nevertheless it may provide an unexpected escape or the opportunity to learn something. Anyway, reasoning by analogy is essential for producing new conclusions that are helpful to solve a problem [3]. Several perspectives make this type of inference a primary issue: (1) in the study of learning, analogies are important in the transfer of knowledge and inferences across different concepts, situations, or domains; (2) analogies are often used in problem solving and reasoning; (3) analogies can serve as mental models to understand new domains; (4) analogy is important in creativity (e.g., it was a frequent mode of thought for such great scientists as Faraday, Maxwell, and Kepler); (5) analogy is used in communication and persuasion; (6) analogy and its cousin, similarity, underlie many other cognitive processes. [3] defined analogy as a partial similarity between different situations that supports further inferences. More precisely, it is a kind of similarity in which the same system of relations holds across different objects. Thus, analogies capture parallelisms across different descriptions (typically, one referring to a past experience and one referring to the situation under consideration). For short, we will call the description coming from prior knowledge, or the domain it is referred to, the *base*, and the description of the current problem, or the domain is referred to, as the *target*.

Typical analogies happen between pairs of specific situations. While this is already a very powerful way of immediately reusing knowledge to solve a problem in a new domain, it may happen that some abstract patterns can be reused across many domains. Since these represent solution schemes whose usefulness has been thoroughly tested, they may be quite promising also in future situations. Albeit not very frequent, these patterns are really valuable and it would be very important to recognize and store them when one comes across them. In practice, this means finding some kind of relevant generalization among different specific analogies. Of course, if one analogy is already a quite complex inference to make, generalization among analogies is much more difficult, because several similar parallelisms must be detected.

This paper proposes the generalization of patterns obtained by analogy, carried out through two main steps: (1) isolating analogous roles of two descriptions coming from different domains, and (2) abstracting from portions of knowledge that have no analogical relationships. The result is a meta-pattern for analogy that generalizes experiences coming from different domains. We propose a multi-strategy approach in which the analogy brings to a generalization, that is, in turn, a novel description to reason over and over again.

The remainder of this paper is organized as follows. In Sect. 2 related works are reviewed and criticized, and preliminary information is provided. Section 3 recalls our procedure to make analogy and subsequent inference. The generalization approach, along with a sample case, is provided in Sect. 4. Finally, Sect. 5 concludes the paper.

2 Preliminaries

According to [15], analogical reasoning involves an inductive step, that hypothesizes the presence of an analogy between two contexts, and a deductive step, that performs truth-preserving reasoning based on the inductively inferred knowledge. However, induction, defined as the process of inferring general knowledge from specific observations, does not fit reasoning by analogy. It would be better described as an *intuition*, since it generates a set of possible mappings that provide analogies with respect to one or more points of view.

Given a point of view, a set of analogical mappings can be used to identify a *recurring meta-pattern* (i.e. a common network of roles). When accomplishing a task, this allows to use the knowledge present in a more familiar base domain to enrich the knowledge of the target domain. The recurring meta-pattern can be seen as an abstract description of the schema shared by the domains under consideration. Unfortunately, the identification of an abstract theory describing several domains is quite complex, since abstractions based on syntactic transformations only might generate an inconsistent set of abstract clauses, even if the ground set is consistent [6].

Most research on analogy operators used formal (Propositional or First-Order) Logic as the most suitable representation for describing high-level cognitive and reasoning processes.

The *Structure Mapping Engine* (SME) [2] uses a local-to-global strategy to structurally align the base and the target, guided by a set of 'programmable' rules. This ensures great flexibility (e.g., allowing to encode similarity or metaphors [2,4]), but requires additional knowledge (e.g., about commutativity). Its formalism [4] involves typed entities that must be declared, and can be translated into ground Horn clauses. It can process 'second order' relations (i.e., relations among relations).

ACME [10] implements a 'cooperative' procedure for parallel satisfaction of a set of interacting constraints[1] represented as a network of supporting and competing hypotheses about what elements to map. The constraints may be of three types: *structural*, satisfied when an exact isomorphism is detected between the analogues; *semantic similarity* supports possible correspondences between elements to the degree that they have similar meaning; *pragmatic centrality* favors correspondences that are pragmatically important to the analogist, either because a particular correspondence between two elements is presumed to hold, or because an element is judged to be sufficiently central that some mapping for it should be found.

---

[1] This connectionist approach to constraint satisfaction was investigated in [16].

LISA [9] builds symbolic representations in neurally inspired computing architectures, an approach named 'symbolic connectionism' [8,11]. Such representations are claimed to give LISA the ability to bind roles to their fillers dynamically (i.e., at need), and to represent the resulting bindings independently of the roles and fillers themselves. LISA provides a natural account of the neural processes, also regarding *working memory* and *long term memory*. The mappings in LISA are sequential, so each one will influence the next; this helps to keep the soundness of the analogies.

DORA [1] performs four basic operations: retrieval of propositions from long-term memory, analogical mapping of propositions currently in working memory, intersection discovery for predication and refinement, and linking of role-filler sets into higher arity structures via self-supervised learning. Just like other works, it assumes that objects and relational roles have a shared pool of basic representational features.

STAR-2 [17] finds analogies by sequentially focusing on parts of the domain. The best mapping for the arguments of the propositions is obtained using parallel computation in a constraint satisfaction network, in order to cope with the explosion of the number of units needed for tensor product representation.

BART [14] performs learning and inference through a Bayesian model. It takes as input vectors representing objects, so that all of a model's relational knowledge must be acquired from non-relational inputs. In this work, the authors focus on learning from positive example only, justifying such a choice as a good approximation of learning in children. Furthermore, since children's learning of relations is often guided by linguistic input from adults, BART focuses on supervised learning using labeled examples.

Finally, Copycat [7] discovers analogies trying to operate in a psychologically realistic way, in a more general framework that simulates fluid concepts and cognitive fluidity. It works on character strings of the form $EFG : MNO = EFH : ?$. The main idea is that high-level cognition-like features emerge from the independent activity of many parallel processes. It is composed by: the Snippet, a kind of archive containing the types of concepts with which the system can work; the Workspace, in which several components cooperate like in a multi-agent system; the Coderack, that provides other agents waiting to be invoked stochastically to carry out sub-tasks into the Workspace, simulating fluidity and creativity.

Only a few works (e.g. DORA) proposed a learning strategy. This is a fundamental issue, since the analogy power resides in the special way in which the cross-domain generalization carries out the inductive step, that represents the quick fix to the lack of experience in the target domain that allows to learn.

3   Analogy and Inference

Our analogical engine can work both in a free setting, trying to find any kind of parallelism between two descriptions, or in a focused setting, where the experimenter can express an initial association between objects in the descriptions and the system must find relevant parallelism involving that association. We refer the interested reader to [12] for details about the underlying algorithms. Here, since we are interested in developing a subsequent phase of processing, we will just briefly recall its main features and general behavior using a running example.

3.1   Representation Formalism

In our approach, each piece of knowledge is formally represented as a Horn clause [13], i.e. a disjunction of literals involving at most one positive literal, where a literal is a (possibly negated) atom. An atom is a predicate applied to its arguments, that are terms (in our case, only constants). A predicate p requiring n arguments is denoted as p/n. Implicitly assuming the inclusive disjunction operator, a clause can also be seen as a set of literals $\{l_0, \neg l_1, \ldots, \neg l_n\}$. The ProLog representation of a clause is $l_0 :- l_1, \ldots, l_n$; where, in the usual interpretation, l_0 is the *head* (i.e., the conclusion of an implication) and $l_1, \ldots, l_n$ is the *body* (i.e., the conjunction of premises of the implication). We may extract the predicate on which an atom $l = p(t_1, \ldots, t_n)$ is built using function predicate$(l) = p/n$, and the terms of l using function terms$(l) = \{t_1, \ldots, t_n\}$. These functions may be straightforwardly extended to literals, while for a clause C we define predicates$(C) = \cup_{l \in C}\{$predicate$(l)\}$ and terms$(C) = \cup_{l \in C}$terms(l).

For our purposes, clauses are not interpreted in the usual way, but they just provide a suitable formalism for expressing the data. The l_i's express properties of, or relationships among, (the objects denoted by) their arguments. Analogical mappings are to be found in the body; the predicate in the head labels the situation that is being described in the body. The heads may be used to provide a preferred focus, i.e. a specific perspective for which the analogical mapping is sought. If exploited, this feature allows us to reduce the search space and direct the operations toward a particular goal of interest. The 'preferred perspective' is enabled when the arity of the predicates in the heads is the same, in which case the system is bound to establish an analogy between corresponding arguments. Using 0-ary predicates in the heads disables this feature. Let us show the formalism with a running example that will be used throughout the paper to illustrate the various steps of the procedure.

Example 1 (Fairy tale and life context). Let us consider the fairy tale *The fox and the grapes* (Walter Crane's version, in *Baby's Own Aesop*, 1887) "This fox has a longing for grapes: he jumps, but the bunch still escapes. So he goes away sour; and, 'tis said, to this hour declares that he's no taste for grapes." would be formalized as:

> *fairy_tale(fox, grape)* :- cannot_reach(fox, grape, fox_does_not_reach_grape),
> wants(fox, grape), is(fox, sour), has(fox, bad_opinion),
> cause(fox_does_not_reach_grape, bad_opinion),
> says(fox, grape_is_no_taste), says(fox, she_is_smart),
> is(grape, not_ripe, grape_is_no_taste).

It means that people tends to belittle things that they would like to obtain but that they cannot, e.g., as in a situation where "John loves Carla but cannot have her, so he spreads a bad opinion about her":

situation(john, carla) :– cannot_have(john, carla, john_cannot_have_carla),
                 loves(john, carla), says(john, carla_is_bad),
                 says(john, he_is_a_charming_man),
                 is(carla, bad, carla_is_bad), feels(john, sad).

The heads in these descriptions suggest that we want to establish an analogy between the *fairy tale* and the given *situation*, in which John plays the role of the fox, and Carla plays the role of the grape. Note that literals *says(fox, she_is_smart)* and *says(john, he_is_a_charming_man)* are useless for the analogy.

Note also that this case needs to represent that the fox has a bad opinion about the grapes because he cannot reach them. But such causal relation, in turn, involves the relation between the fox and the grape (because the former cannot reach the latter), relating it to the bad opinion that the fox has got. Such a situation requires the use of a third argument representing the name of the whole concept for which the fox cannot reach the grape.

The two domains may cross-fertilize each other, providing each other pieces of knowledge that allow to better understand the described situation and help to accomplish tasks (e.g., making comparisons, solving problems, etc.) in them. To build an analogy, an analysis of the relationships in which the objects in the description are involved is fundamental.

3.2 Analogy

In a nutshell, our analogical reasoner initializes the analogy mapping and then progressively expands it, guided by linkedness (i.e., term sharing) among literals. New term or predicate associations are added, as long as they ensure overall consistency of the mapping. The outcome of an analogy is a pair of mappings, one concerning predicates (θ_P) and the other concerning terms (θ_T). Initially, the global predicate mapping is empty ($\theta_P = \emptyset$). Then, if a preferred perspective is expressed by the heads, the mapping of the heads' arguments is taken as a starting point. Suppose that the two clauses C' and C'' input to the procedure have heads l'_0 and l''_0, respectively. If a consistent one-to-one mapping between the arguments of l'_0 and l''_0 exists, then the global term mapping θ_T is initialized to such a mapping. Otherwise, an empty global term mapping $\theta_T = \emptyset$ is initially set, and candidate starting points must be identified by evaluating the shared knowledge between the two descriptions, that provides potential points of contact between them.

Following with the *Fairy tale and life context*, since the heads have the same arity, their terms are mapped, yielding the starting point

$$\theta_T = \{(\text{fox,john}), (\text{grape,carla})\} \qquad \theta_P = \emptyset$$

Then, the expansion phase starts, resulting in the final overall mapping reported in Table 1.

Table 1. Analogy between fairy tale and life context.

Mapped predicates		Mapped terms	
Base clause (fairy tale)	Target clause (life context)	Base clause (fairy tale)	Target clause (life context)
says/2	says/2	fox	john
is/3	is/3	grape	carla
wants/2	loves/2	grape_is_no_taste	carla_is_bad
cannot_reach/3	cannot_have/3	not_ripe	bad
is/2	uses/2	fox_does_not_reach_grape	john_cannot_have_carla
		sour	sad
		she_is_smart	he_is_a_charming_man

3.3 Inference and Re-representation

One-to-one alignment of analogous roles for entities and relationships across domains is of primary importance, because it ensures that part of the structural consistency is verified [2]. For this reason, it is taken as the basis for the inference step, aimed at transferring missing knowledge across the domains. The literals that are completely mapped (i.e., having their predicate and all of their arguments mapped by the analogy), but whose counterpart is not present in the other description, can be immediately projected onto the other domain by taking their analogous counterparts. Moreover, the inference can be extended to partially mapped literals, introducing new names for the missing elements (that represent new knowledge that can be hypothesized in the other domain). We identify these names with the 'skolem_' prefix. Of course, the more Skolem elements in a projection, the less reliable that projection.

In the *Fairy tale and life context* running example, the expected explanation of the phenomenon is: "John has a bad opinion about Carla because he cannot have her love". The inference hypotheses from the fairy tale to the life context are:

> 1: *skolem_cause(john_cannot_have_carla, skolem_bad_opinion)*
> 2: *skolem_has(john, skolem_bad_opinion)*

that fully satisfy the expected interpretation.

4   Analogical Pattern Generalization

The analogical mappings found between pairs of descriptions representing experiences, contexts or concepts can be used to obtain more general analogical schemes. Specifically, each analogy can be 'condensed' in a pattern, that in turn can be used for searching further analogies with other experiences. Such a pattern is stored in the long term memory, allowing self-improvement and growth.

4.1   Formal Definition

Formally, let us consider two atoms having the same arity $n > 0$, $l' = p'(t'_1, \ldots, t'_n)$ and $l'' = p''(t''_1, \ldots, t''_n)$. We define their *atomic analogy* as the pair $a(l', l'') = \langle a_P(l', l''), a_T(l', l'') \rangle$ where: (1) $a_P(l', l'') = \{(p'/n, p''/n)\}$ is the *predicate analogy* between l' and l'', (2) $a_T(l', l'') = \{(t'_1, t''_1), \ldots, (t'_n, t''_n)\}$ is the *term analogy* between l' and l''; if $a_T(l', l'')$ is a one-to-one term mapping. In all other cases it is undefined.

Then, given an analogy $\Theta = \langle \theta_T, \theta_P \rangle$ between two clauses C' and C'', a generalized pattern C with $|C| \leq \min(|C'|, |C''|)$ is outlined as follows. $\forall (l', l'') \in C' \times C''$, $l' = p'(t'_1, ..., t'_n)$, $l'' = p''(t''_1, ..., t''_n)$, for which $\exists a(l', l'')$ s.t. $a_P(l', l'') \subseteq \theta_P \wedge a_T(l', l'') \subseteq \theta_T : l = p(t_1, ..., t_n) \in C$, where: if $p' = p''$, then $p = p' = p''$, otherwise p is a new predicate ($p \notin$ predicates$(C') \cup$ predicates(C'')); $\forall i = 1, \ldots, n$: if $t'_i = t''_i$, then $t_i = t'_i = t''_i$, otherwise t_i is a new term ($t_i \notin$ terms$(C') \cup$ terms(C'')). In other words, if the mapped predicates or terms are the same, their name is preserved, otherwise a new name is generated. For each literal in the pattern, its mappings are stored.

In our running example, the mapping in Table 1 yields the alignment of literals shown in Table 2, from which the following pattern is obtained:

pattern(fairy_tale(fox, grape), situation(john, carla)) :–
  wants_OR_loves(fox_OR_john, grape_OR_carla),
  cannot_reach_OR_cannot_have(fox_OR_john, grape_OR_carla,
        fox_does_not_reach_grape_OR_john_cannot_have_carla),
  says(fox_OR_john, grape_is_no_taste_OR_carla_is_bad),
  says(fox_OR_john, she_is_smart_OR_he_is_a_charming_man),
  is(grape_OR_carla, not_ripe_OR_bad, grape_is_no_taste_OR_carla_is_bad).

where the new predicates and terms have been named by chaining the names of the predicates and terms they generalize, just to let the reader trace back their meaning and role in the original analogy.

As in usual generalization, patterns can be used to find analogies with other descriptions (or even with other patterns). If such analogies do not fully map the

Table 2. Literal mappings between fairy tale and life context.

Fairy tale	Life context
fairy_tale(fox, grape)	*situation(john, carla)*
says(fox, grape_is_no_taste)	says(john, carla_is_bad)
says(fox, she_is_smart)	says(john, he_is_a_charming_man)
is(grape, not_ripe, grape_is_no_taste)	is(carla, bad, carla_is_bad)
wants(fox, grape)	loves(john, carla)
cannot_reach(fox, grape,	cannot_have(john, carla,
fox_does_not_reach_grape)	john_cannot_have_carla)
is(fox,sour)	feels(john,sad)

pattern components, new (and more general) patterns (actually, *meta-patterns*) can be generated. Differently from usual generalization, new patterns do not replace the patterns from which they originated, because each analogy is motivated by a specific perspective, and so the corresponding pattern must be preserved as a representative of that perspective. So, the same (meta-)pattern may give rise to several meta-patterns, based on different combinations of specific domains. This reflects the fact that different aspects of a given situation may have different analogies with other experience according to different perspectives. Note that, as long as (meta-)patterns are progressively generalized (or used with full analogies) with new domains, the surviving elements in the resulting meta-pattern are more and more supported and confirmed, and so they are likely to represent common sense knowledge. In particular, if specific predicate or term names survive many refinements, they are likely to represent fundamental concepts.

The history of a pattern can be traced by recording the origin of each predicate/term in the pattern using 4-tuples of the form:

$$(head, type, pattern\_name, original\_name)$$

where *head* stands for the head of the original clause, *type* indicates if the record concerns a predicate or a term, *pattern_name* represents the name reported in the pattern and *original_name* reports the name in the original clause.

4.2   Evaluation

Traditionally, the evaluation of analogy-related algorithms has been qualitative rather than quantitative. This is mainly because the most interesting thing is whether and how a proposed algorithm can catch relevant parallelisms between descriptions and propose interesting associations. Counting how many successful analogies an algorithm can return can be also quite tricky, since different analogies may be considered as successful depending on the perspective, and thus there is no definite notion of accuracy. As a consequence, no benchmark datasets have been developed on which running experiments. The literature has focused on showing the performance of algorithms on specific relevant cases. We will follow this stream of evaluation.

4.3   Addition and Union

[17] evaluated STAR-2 using a particular case of analogical reasoning among the common properties of addition and set union: associativity, commutativity and the existence of an identity element. The authors emphasize that these properties are isomorphic and involve higher-order propositions, but the two domains do not share common relations or terms. The description of addition contains three predicates *numeric_equality/2* concerning, respectively, associativity, identity and commutativity:

addition :-
   numeric_equality(sum_3_and_4_5, sum_3_4_and_5), sum(number_4, number_5, sum_4_5),
   sum(number_3, sum_4_5, sum_3_and_4_5), sum(number_3, number_4, sum_3_4),
   sum(sum_3_4, number_5, sum_3_4_and_5), numeric_equality(sum_6_0, zero),
   sum(number_6, number_0, sum_6_0), numeric_equality(sum_1_2, sum_2_1),
   sum(number_1, number_2, sum_1_2), sum(number_2, number_1, sum_2_1).

The set-union description contains three predicates *set_equality/2* regarding analogous properties:

union :- set_equality(union_3_and_4_5, union_3_4_and_5), union(set_3, union_4_5, union_3_and_4_5),
       union(set_4, set_5, union_4_5), union(union_3_4, set_5, union_3_4_and_5),
       union(set_3, set_4, union_3_4), set_equality(union_6_ns, null_set),
       union(set_6, null_set, union_6_ns), set_equality(union_1_2, union_2_1),
       union(set_1, set_2, union_1_2), union(set_2, set_1, union_2_1).

Table 3. Addition and set union analogy.

Mapped predicates		Mapped terms			
Base clause	Target clause	Base clause	Target clause	Base clause	Target clause
sum/3	union/3	number_1	set_1	sum_1_2	union_1_2
numeric_equality/2	set_equality/2	number_2	set_2	sum_2_1	union_2_1
		number_3	set_3	sum_3_4	union_3_4
		number_4	set_4	sum_4_5	union_4_5
		number_5	set_5	sum_3_4_and_5	union_3_4_and_5
		number_6	set_6	sum_3_and_4_5	union_3_and_4_5
		zero	null_set	sum_6_0	union_6_ns

The descriptions have no shared knowledge, anyway our approach solved the analogy returning a full mapping between the domains (see Table 3).

4.4   Military and Medical Strategy

[5] adopts a cognitive psychology perspective. To study the process of analogical reasoning in humans, two stories, representing the base and the target domain, are submitted to a group of humans. The humans had to complete the latter based on the former, trying to recognize the knowledge that solves the problem in the target story. Without any suggestion, only 57 % of the subjects provided a complete solution to the analogy, whereas our approach provides directly the correct analogical solution. Let us give details about the stories and the relative expected solution. The base story follows.

A fortress was located in the center of the country. Many roads radiated out from the fortress. A general wanted to capture the fortress with his army. The general wanted to prevent mines on the roads from destroying his army and neighboring villages. As a result the entire army could not attack the fortress along one road. However, the entire army was needed to capture the fortress. So an attack by one small group would not succeed. The general therefore divided his army into several small groups. He positioned the small groups at the heads of different roads. The small groups simultaneously converged on the fortress. In this way the army captured the fortress.

It is translated into the following Horn clause, where each row encodes a sentence in the story.

    conquer(fortress) :–
1.   located(fortress,center), partof(center,country),
2.   radiated(oneroad,fortress), radiated(roads,fortress), partof(oneroad,roads),
3.   capture(general,fortress), use(general,army),
4.   prevent(general,mines), located(mines,oneroad), located(mines,roads),
       destroy(mines,army), destroy(mines,villages),
5.   couldnotuse(army,oneroad),
6.   capture(army,fortress),
7.   couldnotuse(subgroup,oneroad),
8.   splittable(army,subgroups), partof(subgroup,subgroups),
       partof(subgroups,army), destroy(mines,subgroup), notenough(subgroup),
9.   distribute(subgroups,roads),
10.   converge(subgroups,fortress),
11.   capture(subgroups,fortress).

The target story follows.

A tumor was located in the interior of a patient's body. A doctor wanted to destroy the tumor with rays. The doctor wanted to prevent the rays form destroying healthy tissue. As a result the high-intensity rays could not be applied to the tumor along one path. However, high-intensity rays were needed to destroy the tumor. So applying one low-intensity ray would not succeed.

The Horn clause representing the target story is:

    heal(tumor) :–
1.   located(tumor,interior), partof(interior,body),
2.   defeat(doc,tumor), use(doc,rays),
3.   prevent(doc,healthytissue), located(healthytissue,oneslit),
       located(healthytissue,slits), aredestroyed(healthytissue,rays),
4.   couldnotuse(rays,oneslit),
5.   defeat(rays,tumor),
6.   couldnotuse(ray,oneslit),
7. [add] radiated(oneslit,tumor), radiated(slits,tumor), partof(oneslit,slits),
       splittable(rays,ray), partof(ray,subrays), partof(subrays,rays),
       aredestroyed(healthytissue,ray), aredestroyed(healthytissue,subrays),
       notenough(ray).

The last item encodes implicit knowledge, saying that: the cancer can be reached from one or many directions; a slit is one of many slits; the rays can be splitted; and healthy tissue can be damaged and/or destroyed from rays or sub-rays. The expected result must include the missing knowledge of the target story, that is: "The doctor therefore divided the rays into several low-intensity rays. He positioned the low-intensity rays at multiple locations around the patient's body. The low intensity rays simultaneously converged on the tumor. In this way the rays destroyed the tumor." Given the mapping obtained (see Table 4), the inference hypotheses from base to target domain are:

1:	*defeat(subrays,tumor)*
2:	*splittable(rays,subrays)*
3:	*skolem_distribute(subrays,slits)*
4:	*skolem_converge(subrays,tumor)*
5:	*aredestroyed(healthytissue,skolem_villages)*

Hypothesis 1 can be identified as the goal of the problem; it is made up of fully mapped components, making it a conclusive inference. Hypotheses 2, 3 and 4 reveal the procedure to reach the goal; for non-mapped predicates, the *skolem_* prefix was added by the projection step. These inferences are to be considered contingent rather than conclusive. Hypothesis 5 does not make sense (it is a case of fake inference). It is noteworthy that all these statements were absent in the target domain, and have been obtained using the base domain. The consistency with the expected analogical reasoning is evident.

Table 4. Military and medical mapping outcomes

Mapped predicates		Mapped arguments	
Base clause	Target clause	Base clause	Target clause
destroy/2	aredestroyed/2	country	body
capture/2	defeat/2	center	interior
partof/2	partof/2	roads	slits
couldnotuse/2	couldnotuse/2	subgroups	subrays
splittable/2	splittable/2	oneroad	oneslit
use/2	use/2	army	rays
radiated/2	radiated/2	mines	healthytissue
prevent/2	prevent/2	general	doc
located/2	located/2	subgroup	ray
notenough/1	notenough/1	fortress	tumor

4.5   Patterns Assessment

The pattern generation approach was evaluated by carrying on the running example of analogical reasoning between military and medical domains, described in Sect. 3, to a third domain, referred to as *Pharmaceutical*. It tells a story about the purchase of a medicine for which an offertory is needed.

A medicine is located in the warehouse of a pharmacy. A patient needs to purchase the medicine. The patient must face the problem that his money is not enough. As a result the patient cannot purchase the medicine paying the total price. However, the total amount is needed to purchase the medicine. So applying a minor amount cannot succeed.

The clause encoding such concepts is:

get(medicine) :–

1.   located(medicine, warehouse), located(warehouse, pharmacy),
2.   purchase(patient, medicine), use(patient, medicine_total_amount),
3.   mustface(patient, not_enough_money),
4.   couldnotuse(medicine_total_amount, one_money_source),
5.   couldnotuse(partial_amount, one_money_source),
6.   purchase(medicine_total_amount, medicine),
7. [add] splittable(medicine_total_amount, partial_amount),
    partof(partial_amount, many_partial_amounts),
    partof(many_partial_amounts, medicine_total_amount).

Table 5. Pattern and pharmaceutical mapping outcomes

Mapped predicates		Mapped arguments	
Base clause	Target clause	Base clause	Target clause
capture_OR_defeat/2	purchase/2	country_OR_body	pharmacy
partof/2	partof/2	subgroups_OR_subrays	many_partial_amounts
couldnotuse/2	couldnotuse/2	subgroup_OR_ray	partial_amount
located/2	located/2	center_OR_interior	warehouse
use/2	use/2	oneroad_OR_oneslit	one_money_source
prevent/2	mustface/2	army_OR_rays	medicine_total_amount
		mines_OR_healthytissue	not_enough_money
		general_OR_doc	patient
		fortress_OR_tumor	medicine

The mapping is shown in Table 5. Analogous domains can be found in the stored mappings as reported in Table 6. While predicates *use/2, located/2, could-notuse/2* and *partof/2* introduce no novelty, predicates *mustface/2* and *purchase/2* are more interesting because the analogy indicates that *mustface/2* is the 'difficulty' to be overcome, whereas *purchase/2* stands for the main action on which the Pharmaceutical story is built. The term mapping suggestions have no shared knowledge, so each term is traced to different terms in the base domains. For instance, *patient* is traced to the main actors in the other domains (*general* and *doc*); *medicine*, that is the target object in the story, is traced to *fortress* and *tumor*; and so on. As for the analogy between the Military and Medical domains, the gain in the portion of aligned structure before and after analogy was evaluated. The normalized *fs* score is 0.4 for the original clauses, and 0.74 after applying our strategy. So, 34 % more structure has been aligned thanks to the analogy. The evaluation of the inference step is important, as well. Since the projection of knowledge is not empty, the analogy with the Pharmaceutical domain might be represented by an new meta-pattern.

Table 6. Suggestions from original domains

Mapped predicates			Mapped arguments		
Pharmaceutical	Mapping	Source	Pharmaceutical	Mapping	Source
mustface/2	prevent/2	Military	medicine	fortress	Military
		Medical		tumor	Medical
use/2	use/2	Military	patient	general	Military
		Medical		doc	Medical
located/2	located/2	Military	not_enough_money	mines	Military
		Medical		healthytissue	Medical
couldnotuse/2	couldnotuse/2	Military	medicine_total_amount	army	Military
		Medical		rays	Medical
partof/2	partof/2	Military	one_money_source	oneroad	Military
		Medical		oneslit	Medical
purchase/2	capture/2	Military	warehouse	center	Military
	defeat/2	Medical		interior	Medical
			partial_amount	subgroup	Military
				ray	Medical
			many_partial_amounts	subgroups	Military
				subrays	Medical
			pharmacy	country	Military
				body	Medical

5   Conclusions

Analogy is a fundamental inference mechanism to transpose knowledge from known to unknown domains, producing new conclusions that are helpful to solve a problem. In addition to using equal descriptors, as in the mainstream literature, our approach can map also different descriptors that play the same role in the two domains. This paper has shown an approach by which the found analogies can be generalized into meta-patterns, that represent core knowledge and allow further reasoning. This enables more complex reasoning, since by finding an analogy between a meta-pattern and a novel description one may recognize analogies across several different stories, each having a different domain. This can be viewed more in general as a multi-strategy reasoning approach, in which analogies yield generalizations, that in turn are used as novel descriptions to reason over and over again.

Compared to the current literature, our approach allows to learn patterns representing the intuition that leads to a potential solution to the problem, and provides a computational trick that allows to reuse analogies computed in the past. Moreover, it can capture non-syntactic alignments without using meta-descriptions. The challenge to which this approach aims at contributing is the integration of the learned generalizations in the general knowledge network that an agent builds over its lifetime. Such an objective, aimed at overcoming a limit in the current landscape, is not trivial, since *integration* means defining strategies for knowledge addition and retrieval. Even harder difficulties are present in knowledge modification and deletion, since these tasks refer to incremental learning, which is still an open issue.

Future improvements will regard the recognition and mapping of relations with opposite sense. Another interesting direction will be the use of a probabilistic approach to assess the reliability of the mappings. In a multi-strategy perspective, we will study the use of an abductive procedure to check whether the inferred knowledge (mapped or projected) is consistent with the constraints of the target domain, and of an abstraction operator that shifts the representation when needed. Moreover, it will be interesting to allow the analogical reasoner to take advantage of available common sense knowledge, in order to check the soundness of the final result.

References

1. Doumas, L.A.A., Hummel, J.E., Sandhofer, C.M.: A theory of the discovery and predication of relational concepts. Psychol. Rev. **115**(1), 1–43 (2008)
2. Falkenhainer, B., Forbus, K.D., Gentner, D.: The structure-mapping engine: algorithm and examples. Artif. Intell. **41**, 1–63 (1989)
3. Gentner, D.: Analogy. In: A Companion to Cognitive Science, pp. 107–113 (1998)
4. Gentner, D., Markman, A.B.: Structure mapping in analogy and similarity. Am. Psychol. **52**, 45–56 (1997)
5. Gick, M.L., Holyoak, K.J.: Analogical problem solving. Cogn. Psychol. **12**(3), 306–355 (1980)
6. Giordana, A., Saitta, L., Roverso, D.: Abstracting concepts with inverse resolution. In: 8th International Workshop on Machine Learning, pp. 142–146 (1991)
7. Hofstadter, D.R., Mitchell, M.: The copycat project: a model of mental fluidity and analogy-making. In: Advances in Connectionist and Neural Computation Theory. Ablex Publishing Corporation, Norwood (1994)
8. Holyoak, K.J., Hummel, J.E.: The proper treatment of symbols in a connectionist architecture. In: Dietrich, E., Markman, A. (eds.) Cognitive Dynamics: Conceptual and Representational Change in Humans and Machines. Lawrence Erlbaum Associates, Mahwah (2000)
9. Holyoak, K.J., Hummel, J.E.: Understanding analogy within a biological symbol system. In: Holyoak, K.J., Gentner, D., Konikov, B.N. (eds.) The Analogical Mind, pp. 161–195. The MIT Press, Cambridge (2001)
10. Holyoak, K.J., Thagard, P.: Analogical mapping by constraint satisfaction. Cogn. Sci. **13**, 295–355 (1989)
11. Hummel, J.E., Holyoak, K.J.: Distributed representations of structure: a theory of analogical access and mapping. Psychol. Rev. **104**(3), 427–466 (1997)
12. Leuzzi, F., Ferilli, S.: Reasoning by analogy using past experiences. In: Proceedings of the 28th Italian Conference on Computational Logic (CILC 2013), vol. 1068, pp. 115–129. CEUR-WS.org (2013)
13. Lloyd, J.W.: Foundations of Logic Programming, 2nd edn. Springer, Heidelberg (1987)
14. Hongjing, L., Chen, D., Holyoak, K.J.: Bayesian analogy with relational tansformations. Psychol. Rev. **119**(3), 617–648 (2012)

15. Michalski, R.S.: Inferential theory of learning: developing foundations for multistrategy learning. In: Machine Learning: A Multi-strategy Approach, vol. 4, pp. 3–62. Morgan Kaufmann Publishers (1993)
16. Rumelhart, D.E., Smolensky, P., McClelland, J.L., Hinton, G.E.: Schemata and sequential thought processes in PDP models. In: Parallel Distributed Processing, pp. 7–57. MIT Press, Cambridge (1986)
17. Wilson, W.H., Halford, G.S., Gray, B., Phillips, S.: The star-2 model for mapping hierarchically structured analogs. In: The Analogical Mind, pp. 125–159 (2001)

Spectral Features for Audio Based Vehicle Identification

Alicja Wieczorkowska[1], Elżbieta Kubera[2]([✉]), Tomasz Słowik[3],
and Krzysztof Skrzypiec[4]

[1] Polish-Japanese Academy of Information Technology,
Koszykowa 86, 02-008 Warsaw, Poland
alicja@poljap.edu.pl
[2] Department of Applied Mathematics and Computer Science, University of Life
Sciences in Lublin, Akademicka 13, 20-950 Lublin, Poland
elzbieta.kubera@up.lublin.pl
[3] Department of Energetics and Transportation, University of Life Sciences in
Lublin, Akademicka 13, 20-950 Lublin, Poland
tomasz.slowik@up.lublin.pl
[4] Maria Curie-Skłodowska University in Lublin, Pl. Marii Curie-Skłodowskiej 5,
20-031 Lublin, Poland
krzysztof.skrzypiec@poczta.umcs.lublin.pl

Abstract. In this paper we address automatic vehicle identification
based on audio information. Such data are complicated, as they depend
on vehicle type, tires, speed and its change. In our previous research we
designed a feature set for selected vehicle classes, discriminating pairs of
classes. Now, we decided to expand the feature vector and find the best
feature set (mainly based on spectral descriptors), possibly representative
for each investigated vehicle category, which can be applied to a bigger
data set, with more classes. The paper also shows problems related to
vehicles classification, which is detailed in official documents by national
authority for issues related to the national road system, but simplified
for automatic identification purposes. Experiments on audio-based vehicle type identification are presented and conclusions are shown.

Keywords: Intelligent Transport System · Vehicle classification · Audio
signal analysis

1 Introduction

The traffic we experience every day in the roads generates a lot of noise. Many
countries measure this traffic and monitor its density. Such monitoring generates
data (audio, video, etc.) that can be analyzed to estimate how the roads are used,
introduce noise prevention etc. The audio data from the traffic monitoring are
the subject of research presented in this paper. The reason of choosing the audio
is that audio data require less storage space, are cheaper to obtain, and can be
recorded at night or at other low visibility conditions, for instance during bad

M. Ceci et al. (Eds.): NFMCP 2015, LNAI 9607, pp. 163–178, 2016.
DOI: 10.1007/978-3-319-39315-5_11

weather, etc. The audio recorders are easier to install, also in a way that is not visible for drivers, so they are less distractive. In the case of video cameras, the drivers are expecting a radar device, and change their behavior, so audio only recording can be even preferred. Also, audio data can be used to classify vehicles according to the noise produced, into classes approximately uniform with respect to how annoying the noise is. Still, extracting information from the audio data is not simple.

Audio data representing vehicles passing by are very complex, as they depend on many factors. The noise generated by vehicles depends on the vehicle type, speed, traffic intensity, how old the vehicles are, technical parameters, engine type, tires, exhaust system, air intake system, and other factors [12]. If different vehicles have the same type of engine, they sound very similar. On the other hand, the same vehicle sounds different when traveling upwards, downwards, with uniform speed or accelerating/decelerating. Also, the noise generated by old vehicles in very bad condition will be raised by a few dB. Diesel engine is up to 5 dB louder than gasoline engine, whereas electric motor produces very little noise. At very low speed, below 30 km/h, electric motors are hardly audible. This actually is dangerous for pedestrians, as in this case they do not hear the vehicle approaching. At higher speed, tire friction makes these vehicles audible. Also, the road surface is an important factor of vehicle noise, and the difference can be about 5 dB or more.

In order to assess the road traffic in Poland, measurements are performed on various designated roads, at specified dates through the observed year, in day time (6 am–10 pm), at night (10 pm–6 am), and additionally between 8am and 4pm for trucks. The measurements are taken through week days, on Saturdays, Sundays and holidays [8]. Measurements can be done automatically, semi-automatically, or manually. In other countries, data about traffic are also collected. European Union also issued a directive on the framework for the deployment of Intelligent Transport Systems [5], with the purpose (among others) of the facilitation of the electronic data exchange between urban control centers for public or private transport. The United States also prepared a strategic plan for Intelligent Transport Systems (ITS) [11].

1.1 Related Work

The research on audio-based automatic classification of vehicles has already been performed, for varying number of classes. Such research is usually performed for low sampling rate, 8–11.025 kHz, or downsampled for faster processing, and the analyzing window is usually short, 10–50 ms. In our research, we decided to use 48 kHz/24 bit recordings, as this is the standard in modern audio recorders. Also, we decided to use longer analyzing frame, 330ms, to have high resolution spectrum, and longer frames yielded better results in our previous research, with a smaller feature set [14].

Various classifiers have already been applied for audio-based vehicle classification, often with feature selection; extensive literature review on this subject is presented in [7]. In [9], artificial neural network was applied for 3 car classes (and horn as 4th class). Erb [7] applied SVM (support vector machines) and feature selection with linear prediction for 3 classes: car, truck, and van. He obtained 87 % correctness for vehicles traveling at low speed, and 83 % for higher vehicle speeds. For traffic without given probabilities, the best result reached 80 %, and increased to 83 % if class probabilities matched those from the training data [7]. Alexandre et al. in [1] applied multi-layer perceptrons combined with feature selection based on a genetic algorithm, for another 3 classes: car, motorcycle, and truck. Features included mel-frequency cepstral coefficients (MFCC), and zero crossing rate, yielding 93 % correctness for 22 features and 75 % for 66 features [1]. Four target classes were investigated in [15]: bus, car, motor, and truck. The authors used quadratic and linear discriminant analysis, and also k-nearest neighbors method (k-NN) and SVM. Feature vector included, among others, short time energy, average zero crossing rate, and pitch frequency of periodic segments of signals, yielding 80 % correctness for SVM with 12 Mel coefficients [15]. Generally, such research usually aims at recognizing 3–4 classes, for various vehicles, including military ones (see [6]).

1.2 Vehicle Classes

The vehicles can be classified in various ways. In Poland, according to the General Directorate for National Roads and Motorways, the vehicles are classified into the following classes: bicycles, motorcycles (including scooters), cars (including minibuses), vans (light trucks, up to 3.5t), small trucks (above 3.5t), big trucks (above 3.5t with trailers or semitrailers), buses, and tractors (including rollers, excavators etc.) [8]. Detailed specification is also prepared for tax and customs purposes [13]. Modern vehicle classification techniques that can be used for vehicle type recognition are based on data sets of vehicle outlines.

2 Data Collection and Description

Vehicle categories include various types of vehicles, differing in the noise produced; also, similar vehicles can vary in the noise they generate. For instance, scooters differ from motorcycles with respect to the noise generated. Cars include vehicles for up to 9 passengers, including the driver. Off-road vehicle fall into this category, and they can produce more noise if all terrain tires are used. Emergency vehicle also produce different sounds when using audible warning devices. Therefore, to obtain relatively uniform representation of each class, we decided to use most typical vehicles for the following 7 classes:

1. bus,
2. small truck (without trailer),
3. big truck (tractor unit with semitrailer),
4. van,
5. motorcycle, excluding scooters,
6. car,
7. tractor.

Minibuses, scooters, and emergency vehicles using audible warning devices were excluded from our research. Also bicycles were excluded, as they produce almost no sound (we recorded several examples). Tractor units without trailers or semitrailers were also excluded.

Our Data. The audio and video recordings were made in a suburban area near Lublin, Poland, on weekdays in the Fall 2012 (12 November; tractors) and the Spring 2015 (5&10 June; other vehicles), at day time and at good weather, together about 1.5 h of continuous recording. The position of the audio recorder is shown in Fig. 1. The road is approximately flat and straight here. The video was used to mark ground truth data, whereas the audio was used for further investigations. 330 ms frames were taken for parametrization and classification, each frame representing vehicle(s) marked in ground truth labeling; details about amount of frames in each class are given in Sect. 4.2. Our goal was to parameterize the audio data for the automatic recognition of the vehicle type.

Fig. 1. The position of data acquisition

3   Feature Set

Based on our previous research [14], our features are based on 330 ms audio segments (frames), Hamming windowed for FFT (Fast Fourier Transform) spectrum calculation. Most of the features are spectral, plus zero crossing rate - a temporal feature. The feature vector includes standard features used in audio classification, plus additional features designed to discern objects representing our target classes. The features applied are:

- *Audio Spectrum Envelope* - 33 features, SE0, ..., SE32 [17],
- *SUM_SE* - sum of the spectrum envelope values,
- *MAX_SE_V, MAX_SE_IND* - value/index of spectrum envelope maximum,
- *F0_ACor, F0_MLA* - fundamental frequency calculated from the autocorrelation function, and through maximum likelihood algorithm [20],
- *EnAb4kHz* - proportion of the spectral energy above 4 kHz to the entire spectrum energy;
- *Energy* - energy of the entire spectrum;
- *Audio Spectrum Centroid* (SC) - the power weighted average of the frequency bins in the power spectrum. Coefficients were scaled to an octave scale anchored at 1 kHz [17];
- *Audio Spectrum Spread* (SS) - RMS (root mean square) of the deviation of the log frequency power spectrum wrt. *Audio Spectrum Centroid* [17];
- *Zero Crossing Rate* (ZCR) in the time-domain of the sound wave; a zero-crossing is a point where the sign of the function changes;
- *RollOff* - the frequency below which 85 % (experimentally chosen threshold) of the accumulated magnitudes of the spectrum is concentrated,
- *A14, A41, A15, A51, A16, A61, A17, A71, A24, A42, A52, A26, A62, A72, A34, A43, A35, A53, A63, A73, A45, A54, A47, A74, A56, A65, A57, A75, A67, A76* - normalized (with respect to the spectrum energy) energies Axy in the spectral ranges determined in such a way that the energy of this frequency range separates classes x and y, i.e. the class x shows higher energy values than the class y in this range; detailed ranges are shown in Table 1. These ranges were automatically found, using twelve 1-second sounds for each class and 330 ms analyzing frame without overlapping (i.e. 3 frames per second). For the available spectrum resolution, all possible frequency ranges were tested, to find such a range $[R_{low}, R_{up}]$ that the energy in it is between $[min_x, max_x]$ for class x and $[min_y, max_y]$ for class y; if $min_x > max_y$, then $[R_{low}, R_{up}]$ is chosen as the range maximizing the margin between objects of x and y. Margin is calculated as $(min_x - max_y)/(max_x - min_y)$. Not for all pairs of classes such discerning ranges were found;
- *B14, B15, B16, B17, B24, B26, B34, B35, B45, B47, B56, B57, B67* proportion of energies between the indicated spectral ranges, Bxy = Axy/Ayx;
- *BW_10dB, BW_20dB, BW_30dB* - bandwidth of the frequency band comprising the spectrum maximum (in dB scale) and the level drop by 10, 20 and 30 dB, respectively, towards both lower and upper frequencies,
- *f_bus, f_smallTruck, f_bigTruck, f_van, f_motorcycle, f_car, f_tractor* - features discerning a particular class from all other classes, obtained through multiplication of all available Bxy values; the value for the target class should exceed those for other classes (at least this is the case for the data used to determine the frequency ranges Axy).

Altogether, the feature set consists of 97 features. Some of these features were used in our previous research on vehicle classification [14]. New features added in this paper include *Audio Spectrum Envelope, SUM_SE, MAX_SE_V, MAX_SE_IND, F0_ACor, F0_MLA*, BW_10dB, BW_20dB, BW_30dB, and *f_bus*,

Table 1. Spectral ranges Axy: the energy of this frequency range separates classes x and y, i.e. the class x shows higher energy values than the class y in this range

Axy	Lower limit [Hz]	Upper limit [Hz]
A14	90.8203125	723.6328125
A15	17.578125	38.0859375
A16	72.9296875	796.875
A17	26.3671875	46.875
A41	1798.828125	1822.265625
A51	3275.390625	3278.320313
A61	937.5	1054.6875
A71	3369.140625	3418.945313
A24	32.2265625	1116.210938
A26	23.4375	750
A42	1986.328125	2071.289063
A52	117.1875	290.0390625
A62	1001.953125	1297.851563
A72	4209.960938	4212.890625
A34	49.8046875	1183.59375
A35	383.7890625	1127.929688
A43	2554.6875	2589.84375
A53	117.1875	375
A63	4283.203125	4309.570313
A73	3843.75	3849.609375
A45	732.421875	1125
A47	691.40625	1374.023438
A54	117.1875	571.2890625
A74	298.828125	325.1953125
A56	111.328125	541.9921875
A57	87.890625	137.6953125
A65	1069.335938	1397.460938
A75	867.1875	896.484375
A67	770.5078125	1403.320313
A76	316.40625	515.625

f_smallTruck, f_bigTruck, f_van, f_motorcycle, f_car, f_tractor. Features Axy and Bxy were calculated in the same way as in [14] but for 7 classes and for different audio samples (only tractor samples were the same).

The data for experiments were recorded in stereo; the average of both channels was used for calculating features. For each audio frame, the obtained feature vector can be used as input to classifiers, to identify vehicle(s) audible in this frame. Using binary classifiers allows recognition of plural vehicles per segment, i.e. all recognized target classes (for instance, big truck and car).

3.1 Feature Selection

After designing the feature set and performing experiments on this set, we also applied feature selection, as our feature vector is relatively large, so such a procedure is recommended in this case [10]. For each of the classifiers investigated, 3-fold cross validation was applied. We tested 2 versions: with constant number of features to be selected (10 features; number arbitrarily chosen), and with feature importance above a selected threshold (0.5 mean decrease of Gini criterion; threshold arbitrarily chosen, based on the observation of feature importance for all classes). Since better results were obtained in the second case, we decided to choose this feature selection scheme.

4 Experiments

In our experiments we applied SVM, random forests (RF, [3]), and deep learning (DL) architecture (neural network), see Sect. 4.1, using R and packages: h2o, randomForest, and e1071 [16, 18]. In each case, we trained a binary classifier for each target class, to recognize automatically whether a target vehicle sound is present in the analyzed audio data (positive answer of the classifier) or not (negative answer). This is because multiple vehicles can be recorded in the same audio sample, and such samples represent multi-label data. A set of binary classifiers can perform multi-label classification, identifying each vehicle present in the analyzed audio sample.

4.1 Classifiers

SVM looks for a decision surface (hyperplane) that maximizes the margin around the decision boundary. The decision hyperplane should be maximally away from the training data points, called support vectors. Data that is not linearly separable is projected into a higher dimensional space where it is linearly separable. This mapping is done by using kernel functions. In our case, we used kernels in form of radial basis functions (RBF). Such a function has 2 parameters, c and γ, which require tuning for best performance. We applied automatic tuning available in R package (tune.svm).

RF is a set of decision trees, constructed with minimizing bias and correlations between the trees. Each tree is built without pruning to the largest possible extent, using a different N-element bootstrap sample of the N-element training set, i.e. obtained through drawing with replacement. For a K-element feature vector, k features are randomly selected ($k \ll K$, often $k = \sqrt{K}$) for each node of any tree. The best split on these k features is used to split the data in the node, and Gini impurity criterion is minimized to choose the split. The Gini criterion measures of how often an element would be incorrectly labeled if labeled randomly, according to the distribution of labels in the subset. This procedure is repeated M times, to obtain M trees; $M = 500$ in our experiments (standard setting in R). Classification of is performed by simple voting of all trees in RF.

DL architecture is composed of multiple levels of non-linear operations. DL neural network architecture is a multi-layer neural net, with many hidden layers. This algorithm is implemented in h2o as feedforward neural net, with automatic data standardization. Training is performed through back propagation, with adaptive learning. Weights are iteratively updated in so-called epochs, with gridsearch of the parameter space. H2o parameters include large weight penalization and drop-out regularization (ignoring a random fraction of neuron inputs). Standard setting of DL in h2o were used in our experiments.

4.2 Data

The data used in our experiments represented 21 frames (330 ms of audio samples per frame) of positive examples for bus class, 26 for small truck, 39 for big truck, 33 for van, 15 for motorcycle, 33 for car, and 18 for tractor. Negative examples outnumbered the positive for each class, as this reflects the real situation. The data were divided into training and testing part (approximately 2/3 for training and 1/3 for testing), with different vehicles data used for training and for testing, in 3-fold crossvalidation (CV-3). The audio data represented sound of a single vehicle, or multiple vehicles. Positive examples contained sounds of the target class (possibly accompanied with other sounds), and negative examples represented any other classes (single or multiple vehicles), or silence.

4.3 Classification Results

The error and F-measure for our data are shown in Fig. 2. F-measure in 2 cases could not be calculated, as no positive samples were indicated, or precision and recall were both equal to zero. The error is usually small, with the highest error for car classification using SVM, but still much better than random choice.

We used RF to estimate the importance of the proposed features, for one of the folds in CV-3. As examples, we present importance for car, motorcycle and tractor in Fig. 3. MeanDecreaseGini used here is a measure of feature importance based on the Gini impurity index used for the calculation of splits during RF training. When a split of a node is made, Gini index for the two descendent nodes G1 and G2 is less than for the parent node, G0, and the importance I is calculated as $I = G0-G1-G2$. As we can see, our proposed features are of high importance in these cases. Other important features are related to spectral envelope, also in plots not presented in this paper.

We also performed clustering experiments, in order to check how the proposed feature set is grouping vehicle objects [19]. Clustering into 7 clusters was

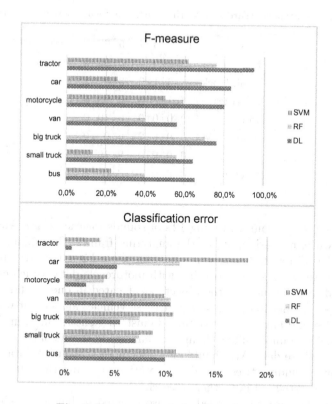

Fig. 2. F-measure and classification error

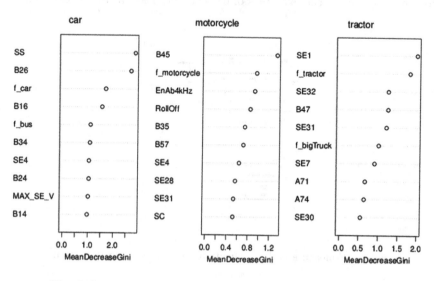

Fig. 3. Importance plot for car, motorcycle, and tractor classes

Table 2. Hierarchical Ward's clustering with Euclidean metrics for our data

cluster no.:	1	2	3	4	5	6	7
bus	12	1	0	0	0	0	0
small truck	3	3	1	0	3	0	3
big truck	4	7	2	0	0	0	0
van	0	0	0	0	0	10	0
motorcycle	0	0	0	2	6	0	0
car	0	0	0	0	0	10	0
tractor	0	0	0	10	8	0	0

performed, for single sounds, taking 12 s of sounds for each class. Analysis was performed without overlapping of the analyzing frames. Exemplary clustering is presented in Table 2. As we can see, data are a bit mixed in clusters. For example, tractor samples are together with motorcycle data, which can be surprising. Still, usually most of the objects are located in one cluster. Only cars and vans are together in one cluster, but these vehicles are similar with respect to produced sound anyway. The obtained clustering shows that our feature set describes vehicle sounds quite well, and explains good results of the classification performed on these data. We should also remember that audio data depend on many factors, including tires, speed, acceleration etc., and these factors can be additionally investigated in further research.

Since our data are imbalanced, we also decided to balance the data. This can be done through downsampling the negative examples for each binary classifiers, or upsampling the examples of each target class [4]. In our experiments,

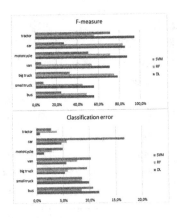

Fig. 4. F-measure and classification error after feature selection

we decided to perform upsampling, i.e. replicating the target class frames. After balancing, equal number of positive and negative examples (audio frames) for each classifier were obtained, i.e. 48–146. Classification and feature selection were performed in 3-fold crossvalidation for each classifier. Features of importance exceeding 0.5 threshold of mean decrease of Gini index were kept in the final feature set. The following features were present in each fold for the target classes:

- bus: SE6–8, SE10–11, SE21–23, SE25, SC, A14, A15, A51, A16, A17, A71, B17, A24, A26, A34, A35, A73, A54, B45, A74, B47, A56, A57, B67, f_bus;
- small truck: SE0, SE4, SE7, SE11, SE14–16, SE30, SUM_SE, MAX_SE_V, Energy, A14, A16, A61, A24, A52, A26, A34, A35, A53, B35, A45, A54, B45, A47, A56, A67, A76;
- big truck: SE0–1, SE3, SE5, SE9–15, SUM_SE, F0_Acor, A14, A16, A61, A24, B24, A52, A26, A34, A35, A53, A63, A45, A54, A47, A74, B47, A56, A57, A75, A67, A76;
- van: SE0, SE7–11, SE13, SE16, SE20, SE23, SE26–32, SUM_SE, F0_Acor, EnAb4kHz, SC, SS, ZCR, RollOff, A14, A41, A15, A16, A61, B16, A17, A24, A52, A26, A62, B26, A34, A43, A53, A54, B45, A74, B47, A56, A65, B56, A67, A76, f_bus, f_smallTruck, f_bigTruck, f_van, f_motorcycle, f_car;
- motorcycle: SE1, SE4, SE6–7, SE13, SE15, SE22, SE25–32, EnAb4kHz, RollOff, B15, A61, A52, B35, A45, B45, A47, B47, A75, f_bigTruck, f_motocycle;

Table 3. Precision and recall of hierarchical classification for our data, using 10 best features in feature selection

Class	DL		RF		SVM	
	precision	recall	precision	recall	precision	recall
HiRot	75.0%	80.0%	58.3%	46.7%	66.7%	40.0%
MedRot	94.1%	97.0%	92.3%	90.9%	93.7%	89.4%
Car	61.4%	85.0%	67.2%	68.3%	56.7%	63.3%
Van	88.9%	77.4%	68.6%	77.4%	65.9%	87.1%
LowRot	81.7%	96.9%	80.9%	78.4%	81.1%	79.4%
Bus+Truck	100.0%	100.0%	66.7%	100.0%	66.7%	100.0%
Bus	94.4%	81.0%	100.0%	71.4%	92.3%	57.1%
Small truck	58.5%	92.3%	67.9%	73.1%	45.2%	53.8%
Big truck	73.5%	92.3%	60.9%	71.8%	62.8%	69.2%
Tractor	100.0%	100.0%	100.0%	66.7%	100.0%	66.7%
Average	82.8%	90.2%	76.3%	74.5%	73.1%	70.6%

- car: SE19–21, SE29–30, SUM_SE, MAX_SE_V, BW_30dB, F0_Acor, SC, SS,
  ZCR, B14, B16, B24, A52, B26, A43, B34, A53, B56, f_smallTruck, f_bigTruck,
  f_van, f_car;
- tractor: SE1, SE7, SE9, SE22–24, SE28–29, SE31–32, A71, B17, A73, B45,
  A74, B47, f_bigTruck, f_motorcycle, f_tractor.

As we can see, the feature designed to identify the target classes, or to discern
between pairs of classes, are of high importance and are kept in the feature set
after the feature selection procedure.

Classification error and F-measure after feature selection for balanced data
are shown in Figure 4. As we can see, classification error decreased after feature
selection in most cases, and deep learning classifiers yield best results.

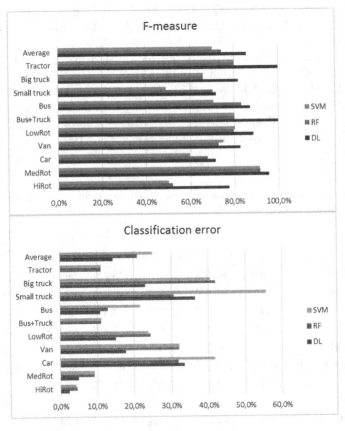

Fig. 5. F-measure and classification error for hierarchical classification, using 10 best
features in feature selection

4.4   Hierarchical Classification

In addition to the experiments described in the previous subsection, we also performed hierarchical classification of our data. Since our experiments are based on audio data, we decided to group the data into supergroups according to criterion that can be heard. Namely, the vehicle classes were grouped into the following 3 groups according to the typical rotational speed of an engine:

- *LowRot*, low rotational speed - tractors, buses, small trucks, big trucks; this class was further subdivided into 2 subclasses:
  - *Bus + Truck*, big trucks, small trucks, and buses - as these vehicles have similar engines,
  - *Tractor*, tractors;
- *MedRot*, medium rotational speed - cars and vans,
- *HiRot*, high rotational speed - motorcycles.

Each of our 7 classes constituted a leaf in this hierarchical classification. The data were balanced through upsampling in the case of unbalanced classes. Feature selection was applied before these experiments, with 2 options as before: for 10 best features, and for features above the threshold (0.5 mean decrease of Gini criterion). The results for 10 best features are shown in Table 3 and Fig. 5, and the results for features above the threshold are shown in Table 4 and Fig. 6. As we can see, better results are obtained again for features above the threshold, and the results are better at higher levels of the hierarchy.

Table 4. Precision and recall of hierarchical classification for our data, using features above the threshold (0.5 mean decrease of Gini criterion) in feature selection

Class	DL		RF		SVM	
	precision	recall	precision	recall	precision	recall
HiRot	82.4%	93.3%	76.9%	66.7%	53.8%	46.7%
MedRot	95.5%	97.0%	92.3%	90.9%	92.4%	92.4%
Car	73.3%	73.3%	66.7%	60.0%	58.9%	55.0%
Van	78.4%	93.5%	67.7%	67.7%	71.4%	80.6%
LowRot	91.6%	89.7%	89.4%	86.6%	90.2%	85.6%
Bus+Truck	100.0%	100.0%	61.1%	100.0%	66.7%	100.0%
Bus	89.5%	81.0%	94.1%	76.2%	69.6%	76.2%
Small truck	64.7%	84.6%	67.9%	73.1%	54.8%	65.4%
Big truck	71.7%	97.4%	63.4%	66.7%	66.7%	71.8%
Tractor	100.0%	100.0%	100.0%	61.1%	100.0%	66.7%
Average	84.7%	91.0%	78.0%	74.9%	72.5%	74.0%

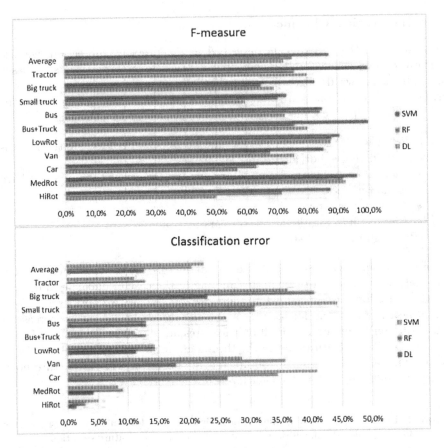

Fig. 6. F-measure and classification error for hierarchical classification, using features above the threshold in feature selection

5   Summary and Conclusions

The features proposed in this paper for audio-based classification of vehicle type yields good results, as the error is below 10–15 % in most cases and improves after feature selection, for 7 classes, which compares favorably with other research, performed for 3–4 classes for similar data. Best results were obtained for deep learning neural network, and hierarchical classification improves the results at higher hierarchy levels. Still, our results can be improved, and we hope to get better results when more data are collected. Also, taking Doppler effect into account may further improve the results, see [2], where data were compared with prerecorded sounds. We can also include subclasses not investigated in this research (scooters, emergency vehicles etc.). Other factors than vehicle type can also be taken into account, including vehicle speed, acceleration, tires, etc. Also, video can be used together with audio data in the vehicle classification task.

Acknowledgments. This work was partially supported by the Research Center of PJAIT, supported by the Ministry of Science and Higher Education in Poland.

References

1. Alexandre, E., Cuadra, L., Salcedo-Sanz, S., Pastor-Sánchez, A., Casanova-Mateo, C.: Hybridizing extreme learning machines and genetic algorithms to select acoustic features in vehicle classification applications. Neurocomputing **152**, 58–68 (2015)
2. Berdnikova, J., Ruuben, T., Kozevnikov, V., Astapov, S.: Acoustic noise pattern detection and identification method in doppler system. Elektronika ir Elektrotechnika **18**(8), 65–68 (2012)
3. Breiman, L.: Random forests. Mach. Learn. **45**, 5–32 (2001). http://www.stat. berkeley.edu/~breiman/RandomForests/cc_papers.htm
4. Chen, C., Liaw, A., Breiman, L.: Using Random Forest to Learn Imbalanced Data. http://statistics.berkeley.edu/sites/default/files/tech-reports/666.pdf
5. Directive 2010/40/Eu of the European Parliament and of the Council of 7 July 2010 on the framework for the deployment of Intelligent Transport Systems in the field of road transport and for interfaceswith other modes of transport. http://eur-lex. europa.eu/LexUriServ/LexUriServ.do?uri=OJ:L:2010:207:0001:0013:EN:PDF
6. Duarte, M.F., Hu, Y.H.: Vehicle classification in distributed sensor networks. J. Parallel Distrib. Comput. **64**, 826–838 (2004)
7. Erb, S.: Classification of Vehicles Based on Acoustic Features. Thesis, Graz University of Technology (2007)
8. General Directorate for National Roads and Motorways (GDDKiA, in Polish). https://www.gddkia.gov.pl/userfiles/articles/z/zarzadzenia-generalnego-dyrektor _13901/zarzadzenie%2038%20Wytyczne%20-%20Zalacznik%20d%20-%20Instruk cja%20%20GPR_2015.pdf
9. George, J., Cyril, A., Koshy, B.I., Mary, L.: Exploring sound signature for vehicle detection and classification using ANN. Int. J. Soft Comput. **4**(2), 29–36 (2013)
10. Hastie, T., Tibshirani, R., Friedman, J.: The Elements of Statistical Learning: Data Mining, Inference, and Prediction. Springer Series in Statistics. Springer, New York (2009)
11. ITS 2015–2019 Strategic Plan. http://www.its.dot.gov/strategicplan.pdf
12. Iwao, K., Yamazaki, I.: A study on the mechanism of tire/road noise. JSAE Rev. **17**, 139–144 (1996)
13. Izba Celna w Przemyślu (the Customs Chamber in Przemyśl, in Polish). http:// www.przemysl.ic.gov.pl/download/sprowadzauto/zas_klasyfikacji_pojazdow_samo. pdf
14. Kubera, E., Wieczorkowska, A., Skrzypiec, K.: Audio-based hierarchic vehicle classification for intelligent transportation systems. In: Esposito, F., Pivert, O., Hacid, M.-S., Rás, Z.W., Ferilli, S. (eds.) ISMIS 2015. LNCS, vol. 9384, pp. 343–352. Springer, Heidelberg (2015). doi:10.1007/978-3-319-25252-0_37
15. Mayvan, A.D., Beheshti, S.A., Masoom, M.H.: Classification of vehicles based on audio signals using quadratic discriminant analysis and high energy feature vectors. Int. J. Soft Comput. **6**, 53–64 (2015)
16. Package 'h2o'. http://cran.r-project.org/web/packages/h2o/h2o.pdf
17. The Moving Picture Experts Group. http://mpeg.chiariglione.org/standards/ mpeg-7

18. The R Foundation. http://www.R-project.org
19. Struyf, A., Hubert, M., Rousseeuw, P.J.: Clustering in an Object-Oriented Environment. http://www.jstatsoft.org/v01/i04/paper
20. Zhang, X., Marasek, K., Raś, Z.W.: Maximum likelihood study for sound pattern separation and recognition. In: 2007 International Conference on Multimedia and Ubiquitous Engineering MUE 2007, pp. 807–812. IEEE (2007)

Probabilistic Frequent Subtree Kernels

Pascal Welke[1]([⊠]), Tamás Horváth[1,2], and Stefan Wrobel[1,2]

[1] Department of Computer Science, University of Bonn, Bonn, Germany
welke@uni-bonn.de
[2] Fraunhofer IAIS, Schloss Birlinghoven, Sankt Augustin, Germany

Abstract. We propose a new probabilistic graph kernel. It is defined
by the set of frequent subtrees generated from a small random sample
of spanning trees of the transaction graphs. In contrast to the ordinary
frequent subgraph kernel it can be computed efficiently for any arbitrary
graphs. Due to its probabilistic nature, the embedding function corresponding to our graph kernel is not always correct. Our empirical results
on artificial and real-world chemical datasets, however, demonstrate that
the graph kernel we propose is much faster than other frequent pattern
based graph kernels, with only marginal loss in predictive accuracy.

1 Introduction

Over the past decade, graph kernels (see, e.g., [7]) have become a well-established
approach in graph mining for their excellent predictive performance. One of the
early graph kernels, the *frequent subgraph kernel* [5], is based on an explicit
embedding of the graphs into a feature space spanned by the set of *all* connected
subgraphs that are frequent w.r.t. the input graph database. It was shown in [5]
that remarkable predictive accuracies can be obtained with this type of graph
kernels on the molecular graphs of small pharmacological compounds.

One of the main drawbacks of frequent subgraph kernels is that the pre-
processing step of generating *all* frequent connected subgraphs is computationally intractable [9]. Many of the practical implementations ignore this complexity
limitation, implying that such systems may become infeasible in practice even
for small datasets. For example, the general-purpose frequent subgraph mining
algorithm FSG [5] requires more than one day to compute all frequent patterns
even for small-sized datasets (50 graphs) of random *sparse* graphs with 25 vertices on average. Approaches that do not disregard the computational limitation
mentioned above resort either to various heuristics for traversing the search space
that result in incomplete output (e.g. [3]) or restrict the input graphs to some
tractable graph class to guarantee both completeness and efficiency (e.g. [10]).

In this work we propose a new approach different from the ones above. We
present a *randomized frequent subtree* kernel that is *not* restricted to any particular graph class and is computable in time *polynomial* in the number of frequent
patterns generated by the algorithm. It follows from the negative complexity
result [9] on mining frequent connected subgraphs from arbitrary graphs that,

© Springer International Publishing Switzerland 2016
M. Ceci et al. (Eds.): NFMCP 2015, LNAI 9607, pp. 179–193, 2016.
DOI: 10.1007/978-3-319-39315-5_12

unless P = NP, such a frequent subgraph kernel can only be achieved by relaxing some of the standard requirements. More precisely, we give up the demand on completeness and calculate a binary feature vector for each graph with the following method: (i) We represent each input graph by a forest formed by k random spanning trees for some *small* k, (ii) compute the set of subtrees frequent in the forest database generated in step (i), and (iii) map the input graphs to the vertices of the Hamming cube in the space spanned by the set of frequent subtrees calculated in (ii). For complexity reasons, a frequent tree pattern is regarded as a subtree of a particular graph in step (iii) only if it is a subtree of the random forest generated for the graph. A random sample of k spanning trees is a forest that can be generated in polynomial time. Combining this with the positive result that frequent subtree mining in forests can be solved with polynomial delay (see, e.g., [2, 10]), we arrive at an algorithm computing the feature vector for *any* graph in time *polynomial* in the combined size of the input database and the set of generated tree patterns.

Regarding steps (i)–(ii), our approach is *sound*, but *incomplete* for two reasons: First, the pattern language is restricted to trees. Second, some frequent subtrees may be missed by the algorithm, as they are not necessarily frequent w.r.t. the random forest generated in step (i). Regarding (iii), this step is sound, but incomplete as well. We still resort to this probabilistic strategy, as it is NP-complete to decide if a tree is subgraph isomorphic to an arbitrary graph. Our somewhat unusual idea is motivated by the intuition that any tree found by our mining algorithm is not only frequent with respect to the database, but with high probability it has a relatively high frequency also in the set of spanning trees for each transaction graph containing it. Thus, there must be a high chance that such a tree pattern will be detected with this method in a query graph as well, if it is part of it.[1] Hence we call our method *probabilistic subtree kernel*. It is significantly different from other techniques commonly called "tree kernels" (see, e.g., [11,12]), as these (i) use homomorphism instead of subgraph isomorphism as the embedding operator and (ii) are not frequency based.

We have empirically evaluated the proposed method on random and on benchmark molecular graph datasets. In particular, we first generated sparse random graphs in the Erdős-Rényi model [6] and investigated the recall of the set of frequent tree patterns for various k (i.e., number of random spanning trees per graph), as well as for different edge factors. (Notice that precision is always 100 % for the soundness of the algorithm.) We found that at least 20 % of *all* frequent subtrees can be recovered from a *single* random spanning tree per graph; for $k = 20$ the recall varies at around 90 %. A similar or even better recall could consistently be observed on different real-world molecular datasets. Not surprisingly, our technique is faster by at least one order of magnitude from an average edge density of around 1.6. In all of our experiments we used the FSG frequent subgraph mining algorithm [5]. From an edge factor of around 2.0, FSG had to be aborted after one day while our algorithm required less than

---

[1] We assume that the query graph has been selected from the same (unknown) probability distribution as the graphs in the input database.

three hours (for frequency threshold 10 %). Using different real-world molecular datasets, in a second step we then compared the predictive performance of our probabilistic approach to that of the ordinary frequent subgraph kernel [5]. We observed only a marginal loss in predictive performance on all datasets. In all of these experiments k was at most 20, in accordance with the recall results on random graphs. We found that with increasing dataset size we needed smaller and smaller values of k to obtain a close approximation of the frequent subgraph kernel's predictive performance. In particular, for the NCI-HIV dataset consisting of more than 40,000 molecular graphs, $k = 5$ sufficed. Putting together, the empirical results suggest that a careful composition of our simple probabilistic technique with some fast (tree) kernel might result in a fast graph kernel of high predictive performance.

The rest of the paper is organized as follows. In Sect. 2 we present our algorithm with some important implementation details. Section 3 describes the empirical evaluation of our approach and Sect. 4 concludes with some interesting questions for further work.

2 The Probabilistic Frequent Subtree Kernel

Several graph kernels have been developed over the past decade for predictive graph mining. A broad range of these graph kernels belong to the class of *convolution kernels* [8]. That is, the input graphs are first decomposed into certain sets of substructures determined by some pattern language and the graph kernels are then defined by the intersection kernel over such sets. Depending on the particular choice of the substructure class (i.e., the pattern language), different graph kernels can be defined in this way. One of the first such graph kernels was defined by means of *frequent connected subgraphs* [5]. That is, the feature space corresponding to the kernel is spanned by the set of connected graphs that occur in at least a certain proportion of the graphs in the input database. The first step of this approach is to generate *all* frequent connected subgraphs, i.e., to solve the following pattern mining problem:

Frequent Connected Subgraph Mining (FCSM) Problem: *Given* a finite set $D \subseteq \mathcal{G}$ for some graph class $\mathcal{G}$ and a threshold $t \in (0, 1]$, *list* the set $F \subseteq \mathcal{P}$ of all pairwise non-isomorphic graphs from some graph class $\mathcal{P}$ that are subgraph isomorphic to at least $\lceil t \cdot |D| \rceil$ graphs in D.

In what follows, $\mathcal{G}$ and $\mathcal{P}$ will be referred to as *transaction* and *pattern* classes. The set F of frequent patterns in D naturally yields a binary vector representation for any arbitrary graph G: We map G to its characteristic vector $\overline{v}_G$ over the universe F, i.e., $\overline{v}_G$ is indexed by F and for all $H \in F$, $\overline{v}_G[H] = 1$ if and only if H is subgraph isomorphic to G. To avoid redundancies in the characteristic vectors over F, the patterns in F are required to be pairwise non-isomorphic.

One of the main limitations of the frequent subgraph kernel [5] lies in the computational intractability of the FCSM problem: If there is no restriction on the transaction graphs in $\mathcal{G}$ and $\mathcal{P}$ consists of all connected graphs of $\mathcal{G}$

then, unless P = NP, the FCSM problem cannot be solved in output polynomial time [9]. As our empirical results of the next section clearly indicate, this negative complexity result makes the frequent subgraph kernel practically infeasible even for small-sized datasets of small sparse graphs. To overcome this limitation, below we propose a probabilistic frequent *subtree* kernel that can be calculated in time polynomial in the combined size of G_1, G_2, and the set of frequent patterns for any *arbitrary* graphs G_1 and G_2.

2.1   Probabilistic Frequent Subtrees

To achieve the goal above, we restrict the pattern language to trees and relax the correctness constraint of the FCSM problem by giving up the requirement of completeness relative to the constrained pattern language of trees. Just restricting the pattern language to trees is, however, not sufficient to get rid of the computational intractability mentioned above; mining frequent trees in arbitrary graphs is not possible in output polynomial time, as otherwise it could solve the Hamiltonian path problem in polynomial time [9]. To overcome this problem, we propose a simple probabilistic approach that proved fast yet powerful enough in all of our empirical experiments. For the sake of simplicity, we assume that the transaction graphs are connected by noting that our algorithm can naturally be generalized to disconnected transaction graphs.

Our approach is very simple: For each transaction graph we first generate a forest formed by a sample of k random spanning trees for some small k, then solve the FCSM problem for this random forest database, and finally use the set of output patterns to define the underlying feature space. With this problem relaxation we arrive at an easy to implement and, as shown in Sect. 3, practically effective frequent subgraph mining algorithm (see Algorithm 1). In addition to the transaction database D and the frequency threshold t given in the definition of the FCSM problem, the input contains an additional parameter $k \in \mathbb{N}$ defining an upper bound on the number of spanning trees to be generated for each transaction graph. The algorithm starts by sampling k spanning trees for each graph in the database. Instead of mining frequent patterns in the input database D directly, we replace each graph G by a forest F_G formed by the vertex disjoint union of the random spanning trees generated for G. This effectively reduces the problem of mining frequent subtrees in arbitrary graph databases D to that of mining frequent subtrees in a database D' consisting of forests. A tree T is regarded to be t-frequent in this setting if and only if it is subgraph isomorphic to at least $\lceil t \cdot |D'| \rceil = \lceil t \cdot |D| \rceil$ forests in D'. As frequent subtree mining in forest databases can be done with polynomial delay [2,10], we arrive at an algorithm that runs in time polynomial in the combined size of D and the set of frequent subtrees in D'.

To distinguish between the output F of the frequent subgraph problem and the output F' of Algorithm 1 on D and t, we will refer to the former set as *frequent patterns* and to the later one as *probabilistic (subtree) patterns* with respect to a threshold t. Clearly, for any D, t, and k, the output of Algorithm 1 is a subset of the set of frequent trees in D, i.e., Algorithm 1 is sound. However,

Given: A graph database $D \subseteq \mathcal{G}$ an integer $k > 0$ and a threshold $t \in (0, 1]$.
Output: A random subset of the set of t-frequent subtrees of D.

1: $D' := \emptyset$
2: **for all** $G \in D$ **do**
3:      Sample k spanning trees of G uniformly at random
4:      Add the forest F_G of those trees up to isomorphism to D'
5: List all t-frequent subgraphs in D'

Algorithm 1: The Probabilistic Subtree Mining Algorithm

it will not necessarily find all frequent patterns, i.e., it is not complete in general. Thus, with this technique, on the one hand we obtain a polynomial time algorithm that is fast for small values of k, on the other hand, however, loose a number of frequent patterns.

A further complexity problem arises when calculating the explicit embedding of a graph into the feature space defined above: Given a set F' of probabilistic tree patterns generated by Algorithm 1 and a graph G, the embedding of G into the feature space spanned by F' cannot be computed in polynomial time (if $P \neq NP$). The reason is that deciding subgraph isomorphism from a tree into an arbitrary graph is NP-complete. Therefore, we allow the embedding to be incorrect and use the following probabilistic embedding based also on a random sample of k spanning trees: Given a tree pattern $T \in F'$ and a graph G,

1. use F_G generated in steps 3–4 of Algorithm 1, if $G \in D$; o/w generate a random forest F_G for G as in steps 3–4 of Algorithm 1,
2. set $\overline{v}_G[T] = 1$ in the binary feature vector $\overline{v}_G$ of G if and only if T is a subgraph of F_G.

Clearly, this embedding is probabilistic because it depends on the randomly generated forest F_G. In the application context of graph kernels, the incompleteness of Algorithm 1 and the incorrectness of the probabilistic embedding sketched above raise two important questions:

1. How stable is the output of Algorithm 1 and what is its recall with respect to *all* frequent subtrees?
2. How does our probabilistic approach influence the predictive performance of the graph kernel obtained?

Regarding the first question, we show in the next section on artificial and real-world chemical graph datasets that (i) the output is very stable even for $k = 1$ and (ii) more than 75 % of the frequent patterns can be recovered by using only ten random spanning trees per graph (i.e., for $k = 10$). The high stability and recall together indicate that the probabilistic embedding of G calculated by the above method has a small Hamming distance to the exact one defined by F.

Regarding the second question, we show on different real-world benchmark graph datasets that our experimental results are comparable with those obtained by the FSG algorithm [5], even for the full set of frequent *subgraphs*. Before presenting these and other empirical results in Sect. 3, we first discuss some implementation issues and analyse the time complexity of Algorithm 1.

2.2 Implementation Issues and Runtime Analysis

Line 3 of Algorithm 1 can be implemented using Wilson's algorithm [13], which has an expected runtime that is linear in the *mean hitting time* of a graph and returns each spanning tree of G with the same probability. It is $\Theta(n^3)$ in the worst case, but conjectured to be much smaller for most graphs [13]. The set of all sampled spanning trees *up to isomorphism* (Line 4) can be computed from the set of sampled spanning trees using some canonical string representation for trees and a prefix tree as data structure (see, e.g., [2] for more details on canonical string representations for labeled graphs). We follow this approach to practically reduce the runtime of the subsequent frequent subtree mining step, as isomorphic spanning trees yield the same subtrees and can safely be omitted. For each tree, this can be done in $O(n \log n)$ time by computing first the tree center and then applying a canonical string algorithm for rooted trees as in [2]. These canonical strings are then stored in and retrieved from a prefix tree in time linear in their size.

Thus, the sampling step of our algorithm runs in expected $O(kn^3)$ time. If we do not require the spanning trees to be drawn uniformly, we can improve on this time and achieve a deterministic $O(km \log n)$ runtime, where m denotes the number of edges. This is achieved by choosing a *random* permutation of the edge set of a graph and then applying Kruskal's minimum spanning tree algorithm using this edge order. It is not difficult to see that this technique can generate random spanning trees with non-uniform probability. As our experimental results on molecular graphs of pharmacological compounds show, non-uniformity has no significant impact on the predictive performance of the graph kernel obtained.

Finally we note that for Line 5, we can use almost any one of the existing algorithms generating frequent connected subgraphs (i.e., subtrees) from forest databases (see, e.g., [2] for an overview on this topic).

3   Experiments

In this section we empirically evaluate our probabilistic approach on artificial and real-world datasets. We start with artificial datasets to study various features of the ordinary and our probabilistic approach on increasingly complex datasets. We then evaluate the two methods on real-world datasets to complement our results on the artificial datasets and to investigate the suitability of our graph kernel for predictive tasks.

In all our experiments, we used FSG [5] in the version provided by the authors[2] to generate frequent subgraphs. In Line 5 of Algorithm 1, we also used FSG to generate frequent subtrees. In this way, we can consistently compare the

---

[2] http://glaros.dtc.umn.edu/gkhome/pafi/overview.

runtimes of the two methods, as none of them is affected by some specific heuristic not used in the other one. However, we expect a significant improvement of our probabilistic method over the traditional one, once a specialized tree mining algorithm is applied. All our experiments were conducted on an Intel i7 CPU with 3.40 GHz and 16GB of RAM.

3.1   Datasets

Any general frequent subgraph mining algorithm is expected to process a broad spectrum of graph databases. Most empirical evaluations, however, concentrate on some particular type of graph data, mostly representing small molecules. These graphs share certain properties, e.g. sparsity, small vertex degree, near planarity, and, in particular, a natural set of frequent patterns corresponding to functional groups. While all these properties (especially the last one) motivate frequent subgraph mining in the first place, it is also important to observe the behavior of a mining technique on data that may or may not have such properties. We therefore conducted experiments on artificial as well as on real-world molecular datasets.

Artificial Datasets. All these datasets consist of unlabeled sparse graphs of varying number of vertices and edges that were generated in the Erdős-Rényi random graph model [6]. The datasets generated are of different structural complexity, where the structural complexity is defined as the *expected edge factor* $q = \frac{m}{n}$ (n is the number of vertices and m the number of edges). For a given q, each graph G in the corresponding dataset is generated as follows: We first draw the number n of vertices uniformly at random between 2 and 50, set the Erdős-Rényi edge probability parameter $p = \frac{2q}{n-1}$, and then generate G on n vertices in the usual way with this p. If the resulting graph is connected, we add it to the dataset.

MUTAG is a dataset of 188 connected compounds labeled according to their mutagenic effect on Salmonella typhimurium. On average, each graph has 20 vertices and 22 edges.

PTC contains 344 connected molecular graphs, labeled according to the carcinogenicity in mice and rats. The graphs have 26 vertices and edges on average.

NCI1, NCI109 consist of 4, 110 (resp. 4127) compounds of which 3530 (resp. 3519) are connected. Both are balanced sets of chemical molecules labeled according to their activity against non-small cell lung cancer (resp. ovarian cancer) cell lines. The average number of vertices is 30, the average number of edges is 32 in both datasets[3].

NCI-HIV consists of 42, 687 compounds of which 39, 337 are connected[4]. The average number of vertices and edges per graph are 41 and 43, respectively.

---

[3] MUTAG, NCI1, NCI109, and PTC were obtained from http://www.di.ens.fr/~shervashidze/code.html.

[4] http://cactus.nci.nih.gov/.

The molecules are annotated with their activity against the human immunodeficiency virus (HIV). In particular, they are labeled by "active" (A), "moderately active" (M), or "inactive" (I). We consider the following three usual binary classification problems: (AMvsI) A and M together versus I, (AvsMI) A versus M and I, and (AvsI) A versus I where instances labeled by M are removed.

ZINC is a subset of $8,946,757$ $(8,946,755$ connected) so called "Lead-Like" molecules from the zinc database of purchasable chemical compounds[5]. The molecules in this subset have a molar mass between $250\,g/mol$ and $350\,g/mol$ and have an average number of vertices and edges 43 and 44, respectively.

3.2   Runtime

In this section, we compare the runtime of FSG and our algorithm (using FSG as the mining subroutine) on artificial datasets and on subsets of the ZINC dataset. We use Wilson's method [13] to generate the random spanning trees and report the combined time for sampling and frequent pattern generation. As already noted, one could use a specialized frequent subtree mining algorithm in combination with our sampling method to further increase the speedup of our method. We experimented with several such publicly available tree mining algorithms but, somewhat surprisingly, they were not able to beat the speed of FSG on the tree dataset.

Figure 1 shows the processing times for expected edge factors (q) varying between 1.0 and 5.0. (Note the log scale used for the y-axis.) We report average execution times over three runs for computing the set of frequent patterns and probabilistic patterns for various numbers of sampled spanning trees (k). It turns out that FSG is very sensible to the parameter q. In order to be able to get any result in reasonable time, we had to restrict the number of graphs in each dataset to 50. Still we had to terminate FSG in several cases where it took more than $24\,h$ ($86,400\,s$), which was consistently the case once q exceeded 1.8. Up to 20 sampled spanning trees, our probabilistic approach is always faster; for $q > 1.4$ it is faster even for $k = 50$ (more precisely, in contrast to FSG, it is able to terminate in significantly less than a day).

Figure 2 shows the time in seconds for our algorithm with $k = 1$ in black and for FSG in gray. It reports results with subsets of the ZINC dataset of increasing size. Though FSG was able to process much larger datasets than in the experiments with artificial datasets, our method always outperformed FSG in runtime on all datasets and for all frequency thresholds. Furthermore, with decreasing frequency threshold, the runtime of our method increases with a much smaller speed. Last but not least, in contrast to our method, FSG has a clear limitation beyond $200,000$ graphs for $t = 5\,\%$ and beyond $100,000$ graphs for $t = 2\,\%$.

---

[5] http://zinc.docking.org/subsets/lead-like.

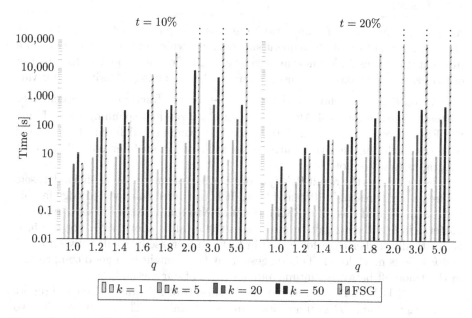

Fig. 1. Runtime (in log-scale) of our method, compared to FSG on artificial datasets of varying expected edge factor q. Dots over bars signal that the run was terminated after 24 h.

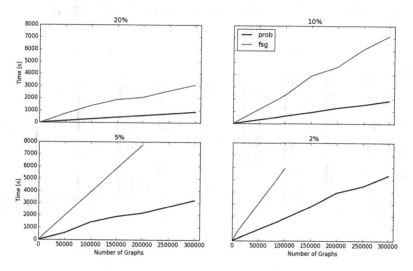

Fig. 2. Runtime results in seconds for our method (black) and FSG (grey), for different frequency thresholds. The x-values show the number of graphs in the subsets of ZINC that were used.

3.3   Recall

As discussed in Sect. 2, for any graph database the pattern set F' found by our algorithm is a subset of all frequent subtrees F_T, which in turn is a subset of all frequent subgraphs F. We now analyze the recall of our method, i.e., the amount of frequent subtree patterns that are found when applying Algorithm 1 for various k and t. To this end, let $R(k,t) := \frac{|F'|}{|F_T|}$ be the fraction of t-frequent tree patterns that are found if Algorithm 1 selects k random spanning trees. Using the FSG algorithm, on each dataset we first compute all frequent connected patterns, including non-tree patterns as well, and then filter out all frequent subgraphs that are not trees.

Figure 3 shows the recall $R(k,t)$ of our method of one run on artificial datasets for frequency thresholds 10 % and 20 %. It is restricted to expected edge factors $q \leq 1.8$, as beyond this value FSG was not able to compute the full set of frequent patterns in less than a day. In most cases the recall is above 40 %. For $k = 5$, it is drastically higher than for $k = 1$; in fact the increase in recall between $k = 5$ and $k = 50$ is much lower. This suggests that $k = 5$ might be a good compromise in the trade-off between runtime and accuracy of our method.

For NCI-HIV and ZINC, we sample 10 subsets of 100 graphs each and report the average value of $R(k,t)$ and its standard deviation. The results on the two datasets can be found in Table 1 for different values of k with frequency thresholds 5 %, 10 %, and 20 %. We have found that at least 95 % of all frequent subgraphs are trees. It can be seen as well that the fraction of the retrieved tree patterns rapidly grows with the number of sampled spanning trees per graph. Sampling 10 spanning trees per graph already results in around 90 % recall for the ZINC dataset and in a recall of around 80 % for the NCI-HIV dataset.

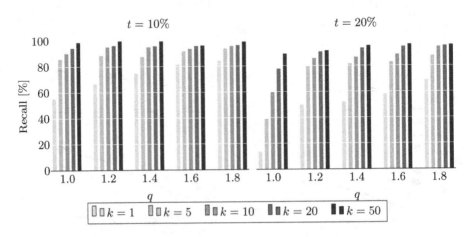

Fig. 3. Recall of our method on artificial graph databases with varying expected edge factor q, for frequency thresholds 10 % and 20 %.

Table 1. Recall with standard deviation of the probabilistic tree patterns on the NCI-HIV and ZINC datasets for frequency thresholds 5 %, 10 %, and 20 %

		$k = 1$	$k = 2$	$k = 3$	$k = 10$	$k = 20$
NCI-HIV	$t = 5\%$	20.13 ± 1.20	35.53 ± 1.34	46.48 ± 0.51	78.32 ± 0.85	91.11 ± 1.29
	$t = 10\%$	20.26 ± 2.22	34.45 ± 1.42	45.40 ± 1.59	79.94 ± 1.82	92.44 ± 1.34
	$t = 20\%$	24.45 ± 1.38	39.76 ± 1.68	50.41 ± 1.14	83.38 ± 1.40	94.72 ± 1.31
ZINC	$t = 5\%$	36.80 ± 0.87	56.70 ± 1.65	68.42 ± 0.94	92.50 ± 0.45	97.92 ± 0.55
	$t = 10\%$	32.77 ± 1.89	51.36 ± 1.84	64.47 ± 1.40	92.49 ± 1.18	86.70 ± 22.83
	$t = 20\%$	31.03 ± 2.59	48.99 ± 3.05	61.41 ± 3.41	90.53 ± 1.28	97.89 ± 0.40

3.4   Stability of Probabilistic Subtree Patterns

The results of Sect. 3.3 above indicate that a relatively high recall of the frequent tree patterns can be achieved on molecular graphs and on artificial datasets, even for a very small number of random spanning trees. In this section we report empirical results showing that the output pattern set of Algorithm 1 is quite stable (i.e., independent runs of our probabilistic tree mining yield similar sets of frequent patterns). To empirically demonstrate this advantageous property, we ran Algorithm 1 several times on the same values of the parameters k and t and observed how the union of the probabilistic tree patterns grows.

To this end, we fix two sets of graphs, each of size approximately $40,000$, as follows: We take all connected graphs in NCI-HIV, as well as a random subset $ZINC_{40k}$ of ZINC that contains $40,000$ graphs. We run Algorithm 1 10 times for the datasets obtained with parameters $k = 1$ and $t = 10\%$. Each execution results in a set F_i' of probabilistic subtree patterns, from which we define $U_i = \bigcup_{j=0}^{i} F_j'$ with $F_0' := \emptyset$. Table 2 reports $|F_i' \setminus U_{i-1}|$, i.e., the number of *new* probabilistic subtree patterns found in iteration i for $i = 1, \ldots, 10$ on the left. For an initial number of $3,920$ (NCI-HIV) and $9,898$ ($ZINC_{40k}$) probabilistic patterns, the number of newly discovered patterns drops to at most 22 for the following iterations.

We observed a similar behavior consistently on artificial graphs as well (over all observed edge factors, all numbers of sampled spanning trees, and all frequency thresholds). Due to lack of space, Table 2 only shows the results for $t = 10\%$, $k = 10$, and 5 iterations on the right. As in the evaluation in Sect. 3.3, each artificial dataset consists of 50 graphs.

These results together suggest that the generated feature set does *not* depend too much on the particular spanning trees selected at random. Overall, we have found that independent runs of our algorithm yield similar feature sets on the same data. This observation, combined with the remarkable recall results of the previous section, is essential for our kernel; high recall and stability together indicate that the predictive performance of the (computationally intractable) *exact* frequent subtree kernel can closely be approximated by our (computationally feasible) *probabilistic* frequent subtree kernel even for small values of k.

Table 2. Repetitions of the experiment with $t = 10\%$ and $k = 1$ sampled trees on NCI-HIV and ZINC (left), and $k = 10$ for random graphs with different edge factors q (right). The numbers reported are the number of probabilistic patterns that were not in the union of all probabilistic patterns found up to the current iteration.

Iteration	1	2	3	4	5
$q = 1.0$	692	2	5	8	3
$q = 1.2$	750	2	0	0	11
$q = 1.4$	806	18	0	0	0
$q = 1.6$	824	1	0	0	0
$q = 1.8$	824	2	0	0	0
$q = 2.0$	850	0	0	1	0
$q = 3.0$	814	26	1	4	0
$q = 5.0$	822	4	0	0	20

Iteration	1	2	3	4	5	6	7	8	9	10
NCI-HIV	3920	20	5	10	14	7	2	6	7	2
$ZINC_{40k}$	9898	18	17	11	10	22	7	7	9	1

3.5 Predictive Performance

In this section we show that the predictive performance of the probabilistic subtree kernel compares favorably with that of the frequent subgraph kernel. We deliberately consider the more expressive complete output of FSG, including also cyclic patterns, because we compare the runtime of our method to that of FSG needed to compute *all* frequent subgraphs. We choose, as does most related work, a wrapper method and report the achieved area under the ROC-curve (AUC) of a well trained support vector machine (SVM) [4]. To this end, we consider the seven binary classification problems described in Sect. 3.1. We compare the predictive performance of (i) the frequent subgraph kernel computed by FSG [5] with that of (ii) the probabilistic frequent subtree kernel for different k and for different frequency thresholds. For (ii), we use only the results with Wilson's random spanning tree sampling algorithm [13]; we obtained nearly identical accuracy and runtime results with the greedy sampling algorithm based on Kruskal's method. For our evaluation, we use the SVM provided by the libSVM package [1] with a radial basis function kernel.

We repeat Algorithm 1 four times using different sets of sampled trees and report the average and standard deviation of AUC values from a 3-fold cross validation for each resulting feature set. The same procedure is applied to the frequent subgraph pattern set, here we use a different splitting for the cross validation in each run.

Table 3 shows the results for the classification problems on MUTAG, NCI1, NCI109, and PTC. We can see that the frequent subgraph kernel outperforms our probabilistic subtree kernels for all frequency thresholds and all choices of k. However, if we select $k = 20$ spanning trees, the accuracy is fairly close to that of the exact frequent subtree kernel for all datasets and for all frequency thresholds. Furthermore, the results suggest that we can achieve or perhaps even increase the predictive accuracy of the exact frequent subgraph kernel at a certain frequency threshold t by using the probabilistic frequent subtree kernel with parameters $k = 20$ and frequency threshold $t/2$. It is also worth noting that the increase of accuracy slows down in the function of k; the gain of moving from 1 to 5

Table 3. AUC values [%] of an SVM classifier on MUTAG, NCI1, NCI109, and PTC for frequency thresholds t between 1 % and 20 % when using features generated by FSG and our method for $k \in \{1, 2, 5, 10, 20\}$.

t	k	MUTAG	PTC	NCI1	NCI109
1 %	1	81.72 ± 1.22	56.20 ± 1.54	79.73 ± 0.26	78.64 ± 0.20
1 %	2	82.98 ± 0.46	57.03 ± 0.88	81.74 ± 0.22	80.89 ± 0.15
1 %	5	85.47 ± 0.80	59.18 ± 0.54	83.45 ± 0.12	83.07 ± 0.14
1 %	10	88.33 ± 0.30	59.67 ± 0.26	84.09 ± 0.10	83.79 ± 0.15
1 %	20	89.32 ± 0.14	60.10 ± 0.09	84.43 ± 0.06	84.23 ± 0.05
1 %	FSG	91.18 ± 0.46	63.62 ± 1.01	86.87 ± 0.10	86.84 ± 0.09
5 %	1	80.79 ± 1.26	54.92 ± 1.69	76.90 ± 0.40	75.67 ± 0.23
5 %	2	82.30 ± 0.41	55.05 ± 1.25	78.87 ± 0.17	77.73 ± 0.17
5 %	5	84.20 ± 0.90	56.12 ± 0.67	80.75 ± 0.17	80.31 ± 0.16
5 %	10	86.35 ± 0.15	56.14 ± 0.29	81.60 ± 0.10	81.12 ± 0.13
5 %	20	87.66 ± 0.26	56.34 ± 0.19	82.15 ± 0.05	81.73 ± 0.05
5 %	FSG	89.01 ± 0.64	58.00 ± 1.86	83.76 ± 0.13	83.86 ± 0.06
10 %	1	80.99 ± 1.23	54.05 ± 1.84	75.41 ± 0.43	74.10 ± 0.28
10 %	2	82.60 ± 0.44	54.35 ± 1.48	77.28 ± 0.22	76.08 ± 0.17
10 %	5	84.22 ± 0.86	54.17 ± 0.87	79.09 ± 0.16	78.05 ± 0.14
10 %	10	86.23 ± 0.16	53.94 ± 0.28	79.95 ± 0.09	79.01 ± 0.10
10 %	20	86.95 ± 0.11	53.99 ± 0.19	80.44 ± 0.05	79.61 ± 0.07
10 %	FSG	87.34 ± 0.46	56.76 ± 1.96	81.66 ± 0.10	81.55 ± 0.24
20 %	1	81.02 ± 1.43	53.36 ± 2.16	72.78 ± 0.35	70.84 ± 0.32
20 %	2	83.12 ± 0.53	53.05 ± 0.79	74.94 ± 0.22	73.77 ± 0.17
20 %	5	84.68 ± 0.82	52.34 ± 0.89	77.05 ± 0.15	76.13 ± 0.11
20 %	10	86.92 ± 0.16	51.86 ± 0.52	77.79 ± 0.06	76.90 ± 0.10
20 %	20	88.10 ± 0.06	51.97 ± 0.22	78.15 ± 0.06	77.33 ± 0.08
20 %	FSG	88.36 ± 0.00	55.82 ± 2.59	77.41 ± 0.09	77.92 ± 0.02

spanning trees is much larger than that from 5 to 10 on all datasets except MUTAG, where the second increase is comparable to the first. We assume that this behavior on MUTAG is due to the small number of molecules in the dataset.

The results on NCI-HIV are presented in Table 4. On the one hand, one can see that from a frequency threshold of 10 %, the results with the frequent subgraph kernel are more stable than those with the probabilistic frequent subtree kernel on all three problems. Though the frequent subgraph kernel outperforms the probabilistic frequent subtree kernel on the same frequency threshold, the difference seems marginal once we compare the best results on each problem, especially in light of the runtime benefits presented above. On the other hand, however, for the frequent subgraph kernel, the results could be calculated only for $t = 10\,\%$,

Table 4. Average AUC values for the three learning problems on the NCI-HIV benchmark dataset for the frequent subgraph kernel and the probabilistic frequent subtree kernel for $k = 1, 2$ and for different frequency thresholds.

t	k	AvsI	AMvsI	AvsMI
5 %	FSG	o.o.m	o.o.m	o.o.m
5 %	1	89.27 ± 0.20	72.35 ± 0.23	88.23 ± 0.24
5 %	2	89.94 ± 0.12	74.09 ± 0.69	89.09 ± 0.74
5 %	5	91.17 ± 0.13	75.65 ± 0.27	90.63 ± 0.17
10 %	FSG	91.31 ± 0.38	75.29 ± 0.24	90.82 ± 0.31
10 %	1	88.53 ± 0.81	71.32 ± 0.54	87.45 ± 1.18
10 %	2	88.28 ± 1.51	71.09 ± 0.21	87.29 ± 0.62
10 %	5	91.11 ± 0.23	74.30 ± 0.18	90.27 ± 0.08
20 %	FSG	91.35 ± 0.39	74.24 ± 0.26	90.57 ± 0.17
20 %	1	86.75 ± 0.76	68.55 ± 0.73	86.00 ± 0.74
20 %	2	86.40 ± 1.00	68.79 ± 0.61	85.79 ± 0.74
20 %	5	90.29 ± 0.28	73.17 ± 0.56	90.27 ± 0.53

while for the probabilistic frequent subtree kernel we obtained the result in half of the time for $t = 5\,\%$. For this frequency threshold, FSG was unable to produce any result because it ran out of memory. For larger frequency thresholds, we had difficulties with training the SVM using all frequent patterns because of its excessive memory usage. These observations clearly show the limitation of the frequent subgraph kernel over the probabilistic frequent subtree kernel when the predictive performance required can be achieved only for low frequency thresholds. Finally we note that there is no improvement when sampling two instead of one spanning tree per graph, but a drastic increase when increasing k to 5. This result fits well with the evaluation of our method on the artificial datasets.

4   Conclusion and Future Work

We have presented a kernel for graph structured data that is based on probabilistic subtree patterns, i.e., on frequent subtrees in a forest database obtained by randomly selecting k spanning trees for each transaction graph in the input database, for some small value of k. Our empirical results on various random graph datasets generated in the Erdős-Rényi model show that even for small values of k ($k \leq 20$), the output of the probabilistic frequent subtree mining algorithm is of high recall and stability. This implies that the predictive performance of the corresponding probabilistic frequent subtree kernel must closely approximate to that of the exact frequent subtree kernel. Runtime comparisons with the FSG frequent subgraph mining algorithm clearly demonstrate the superiority of our probabilistic approach; the speed of the probabilistic frequent subtree mining algorithm is faster by at least one order of magnitude.

Our empirical results on various real-world benchmark graph datasets show that the probabilistic feature space considered is expressive enough in terms of predictive performance compared to that of the ordinary frequent subgraph kernel. Furthermore, our graph kernel is not only faster than the frequent subgraph kernel, but has a much smaller memory footprint in all stages.

We are currently working on some formal properties of the proposed method, e.g., on probabilistic guarantees for (t_1, t_2)-frequent subtrees where t_1 is a frequency threshold for a tree pattern within the set of spanning trees of a graph, whereas t_2 within the database. These results will then be turned into an algorithm that, for a given confidence value δ specified by the user, generates each (t_1, t_2)-frequent subtree with probability at least δ. We are also considering the design and implementation of a frequent subtree mining algorithm for unlabeled free trees that is able to effectively process massive forest transaction datasets.

One of the strengths of our method is that it is not restricted to any particular graph class. This advantageous property allows us to empirically investigate the proposed graph kernel on more complicated graph classes beyond molecular graphs, such as the k-neighborhood graphs of the web graph or RDF graphs.

References

1. Chang, C.-C., Lin, C.-J.: Libsvm: a library for support vector machines. ACM Trans. Intell. Syst. Technol. (TIST) **2**(3), 27 (2011)
2. Chi, Y., Muntz, R.R., Nijssen, S., Kok, J.N.: Frequent subtree mining - an overview. Fundam. Inform. **66**(1–2), 161–198 (2001)
3. Cook, D.J., Holder, L.B.: Substructure discovery using minimum description length and background knowledge. J. Artif. Intell. Res. (JAIR) **1**, 231–255 (1994)
4. Cortes, C., Vapnik, V.: Support-vector networks. Mach. Learn. **20**(3), 273–297 (1995)
5. Deshpande, M., Kuramochi, M., Wale, N., Karypis, G.: Frequent substructure-based approaches for classifying chemical compounds. IEEE Trans. Knowl. Data Eng. **17**(8), 1036–1050 (2005)
6. Erdős, P., Rényi, A.: On random graphs. Publicationes Mathematicae **6**(5), 290–297 (1959)
7. Gärtner, T.: Kernels for structured data. Ph.D. thesis, Universität Bonn (2005)
8. Haussler, D.: Convolution kernels on discrete structures. Technical report UCSC-CRL-99-10, Univerisity of California - Santa Cruz, July 1999
9. Horváth, T., Bringmann, B., De Raedt, L.: Frequent hypergraph mining. In: Muggleton, S.H., Otero, R., Tamaddoni-Nezhad, A. (eds.) ILP 2006. LNCS (LNAI), vol. 4455, pp. 244–259. Springer, Heidelberg (2007)
10. Horváth, T., Ramon, J.: Efficient frequent connected subgraph mining in graphs of bounded tree-width. Theor. Comput. Sci. **411**(31–33), 2784–2797 (2010)
11. Mahé, P., Vert, J.-P.: Graph kernels based on tree patterns for molecules. Mach. Learn. **75**(1), 3–35 (2009)
12. Shervashidze, N., Schweitzer, P., Van Leeuwen, E.J., Mehlhorn, K., Borgwardt, K.M.: Weisfeiler-lehman graph kernels. J. Mach. Learn. Res. **12**, 2539–2561 (2011)
13. Wilson, D.B.: Generating random spanning trees more quickly than the cover time. In: Proceedings of the Twenty-Eighth Annual ACM Symposium on Theory of Computing, pp. 296–303. ACM (1996)

Heterogeneous Network Decomposition and Weighting with Text Mining Heuristics

Jan Kralj[1,2]([✉]), Marko Robnik-Šikonja[3], and Nada Lavrač[1,2,4]

[1] Jožef Stefan Institute, Jamova 39, 1000 Ljubljana, Slovenia
nada.lavrac@ijs.si
[2] Jožef Stefan International Postgraduate School,
Jamova 39, 1000 Ljubljana, Slovenia
jan.kralj@ijs.si
[3] Faculty of Computer and Information Science,
Večna pot 113, 1000 Ljubljana, Slovenia
marko.robnik@fri.uni-lj.si
[4] University of Nova Gorica, Vipavska 13, 5000 Nova Gorica, Slovenia

Abstract. The paper presents an approach to mining heterogeneous information networks applied to a task of categorizing customers linked in a heterogeneous network of products, categories and customers. We propose a two step methodology to classify the customers. In the first step, the heterogeneous network is decomposed into several homogeneous networks using different connecting nodes. Similarly to the construction of bag-of-words vectors in text mining, we assign larger weights to more important nodes. In the second step, the resulting homogeneous networks are used to classify data either by network propositionalization or label propagation. Because the data set is highly imbalanced we adapt the label propagation algorithm to handle imbalanced data. We perform a series of experiments and compare different heuristics used in the first step of the methodology, as well as different classifiers which can be used in the second step of the methodology.

Keywords: Network analysis · Heterogeneous information networks · Network decomposition · PageRank · Text mining heuristics · Centroid classifier · SVM

The paper presents an initial investigation into merging semantic data mining techniques with network analysis technologies. Semantic data mining is a research field that allows us to create deep and meaningful explanations of data. The downside of SDM methods is their high computational cost. We aim mitigate this by constructing a new methodology that will use network analysis to identify interesting parts of a network and allow us to decrease the amount of background knowledge searched by SDM methods. In this paper, we present results of a network analysis study of the data. We analyzed networks, arising from the data set, using community detection and network ranking algorithms. We also present an initial attempt at using SDM methods to discover rules.

© Springer International Publishing Switzerland 2016
M. Ceci et al. (Eds.): NFMCP 2015, LNAI 9607, pp. 194–208, 2016.
DOI: 10.1007/978-3-319-39315-5_13

1   Introduction

The field of *network analysis* is well established and exists as an independent research discipline since the late seventies [24] and early eighties [1]. In recent years, analysis of *heterogeneous information networks* [20] has gained popularity. In contrast to standard (homogeneous) information networks, heterogeneous networks describe heterogeneous types of entities and different types of relations.

This paper addresses the task of mining *heterogeneous information networks* [20]. In particular, for a user-defined node type, we use the method of classifying network nodes through network decomposition; this results in homogeneous networks whose links are derived from the original network. Following [6], the method constructs homogeneous networks whose links are weighed with the number of intermediary nodes, connecting two nodes. After the individual homogeneous networks are constructed, we consider two approaches for classification of network nodes. We classify the nodes either through label propagation [25], or using a propositionalization approach [6]. The latter allows the use of standard classifiers such as the centroid and SVM classifier on the derived feature vector representation. The propositionalization approach was already applied to a large heterogeneous network of scientific papers from the field of psychology in our previous work [13]. In this work, we propose two improvements to the presented methodology: (1) a new variant of label propagation is proposed, resulting in improved classification performance on imbalanced data sets; (2) new heuristics for homogeneous network construction are proposed, which are inspired by word weighting heuristics used in text mining and information retrieval.

The paper is structured as follows. Section 2 presents the related work. Section 3 presents the two-stage methodology for classification in heterogeneous networks. We present a method for constructing homogeneous networks from heterogeneous networks and two methods for classification of nodes in a network: a network propositionalization technique and a label propagation algorithm. Section 4 presents how these two methods are further improved. We first introduce a variant of the label propagation algorithm, which improves performance on imbalanced data sets. We then show how the homogeneous network construction can be improved by using different weighting heuristics. Section 5 presents the application of the methodology on a challenge data set of customers linked to products they purchased, followed by the three stage analysis of the results. First, we examine the effect of using different classifiers in the final step of classification via propositionalization. Second, we test different heuristics for homogeneous network construction. Finally, we analyze the improved label propagation method for imbalanced data sets. Section 6 concludes the paper and presents the plans for further work.

2   Related Work

In network data analysis, instances are connected in a network of connections. In ranking methods like Hubs and Authorities (HITS) [11], PageRank [18], SimRank [9] and diffusion kernels [12], authority is propagated via network edges to discover high ranking nodes in the network. Sun and Han [20] introduced the concept of *authority* ranking for heterogeneous networks with two node types (bipartite networks) to simultaneously rank nodes of both types. Sun et al. [21] address authority ranking of all nodes types in heterogeneous networks with a star network schema, while Grčar et al. [6] apply the PageRank algorithm to find PageRank values of only particular type of nodes.

In network classification, a typical task is to find class labels for some of the nodes in the network using known class labels of the remaining network nodes. A common approach is propagation of labels in the network, a concept used in [23,25]. An alternative approach is classification of network nodes through propositionalization, described in [6]. There, a heterogeneous network is decomposed into several homogeneous networks which are used to create feature vectors corresponding to nodes in the network. The feature vectors are classified by SVM [4,14,16], kNN [22] or centroid classifier [7] to predict class values of these nodes. The network propositionalization approach was also used in [13].

Our work is also related to text mining, specifically to bag-of-words vector construction. Here it is important to correctly set weights of terms in documents. Simple methods like term frequency are rarely used, as the term-frequency inverse-document-frequency (tf-idf) weighting introduced in [10] is more efficient. A number of weighting heuristics which also take into account labels of documents have been proposed, such as the χ^2, information gain [3], Δ-idf [17], and relevance frequency [15].

3   Methodology

This section addresses the problem of mining heterogeneous information networks, as defined by Sun and Han [20], in which a certain type of nodes (called the target type) is labeled. A two step methodology to mine class labeled heterogeneous information networks is presented. In the first step of the methodology, the heterogeneous network is decomposed into a set of homogeneous networks. In the second step, the homogeneous networks are used to predict the labels of target nodes.

3.1   Network Decomposition

The original heterogeneous information network is first decomposed into a set of homogeneous networks, containing only the target nodes of the original network. In each homogeneous network two nodes are connected if they share a particular direct or indirect link in the original heterogeneous network. Take as an example a network containing two types of nodes, *Papers* and *Authors*, and two edge types, *Cites* (linking papers to papers) and *Written_by* (linking papers to authors). From it, we can construct two homogeneous networks of papers: the first, in which two papers are connected if one paper cites another, and the second, in which they are connected if they share a common author (shown in Fig. 1). The choice of links used in the network decomposition step

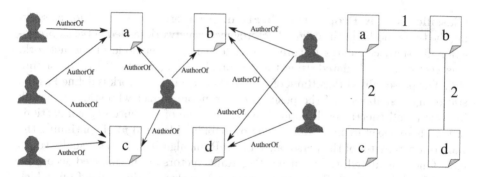

Fig. 1. An example of a heterogeneous network, decomposed into a homogeneous network where papers are connected if they share a common author. Weights of the edges are equal to the number of authors that contributed to both papers

requires expert who takes the meaning of links into account and chooses only the decompositions relevant for a given task.

3.2 Classification

In the second step of the methodology, the homogeneous networks are used to classify the nodes. We compare two approaches to this task: the label propagation algorithm [25] and the network propositionalization approach [6].

Label Propagation. The label propagation algorithm starts with a network adjacency matrix $M \in \mathbb{R}^{n,n}$ and a class matrix $Y \in \mathbb{R}^{n,|C|}$, where $C = \{c_1, \ldots, c_m\}$ is the set of classes, with which the network nodes are labeled. The j-th column of Y represents the j-th label of C, meaning that Y_{ij} is equal to 1 if the i-th node belongs to the j-th class and 0 otherwise. The algorithm constructs the matrix $S = D^{-\frac{1}{2}} M D^{-\frac{1}{2}}$, where D is a diagonal matrix and the value of each diagonal element is the sum of the corresponding row of M. The algorithm iteratively computes $F(t) = \alpha S F(t-1) + (1-\alpha)Y$ until there are no changes in the matrix $F(t)$. The resulting matrix F is used to predict the class labels of all unlabeled nodes in the network. Zhou et al. [25] show that the iteration converges to the same value regardless of the starting point $F(0)$. They also show that the value F^* that $F(t)$ converges to can also be calculated by solving a system of linear equations, as

$$F^* = (I - \alpha S)^{-1} Y. \tag{1}$$

To classify a heterogeneous network, decomposed into k homogeneous networks, we propose classification of nodes using *all* available connections from all k homogeneous network. We construct a new network with the same set of nodes as in the original homogeneous networks. The weight of a link between two nodes is calculated as the sum of link weights in all homogeneous networks. In effect, if the original networks are represented by adjacency matrices $M_1, M_2, \ldots, M_k$, the new network's adjacency matrix equals $M_1 + M_2 + \cdots + M_k$.

Classification by Propositionalization. An alternative method for classifying the target nodes in the original heterogeneous network (called network propositionalization) calculates feature vectors for each target node in the network. The vectors are calculated using the personalized PageRank (P-PR) algorithm [18]. The personalized PageRank of node v (P-PR$_v$) in a network is defined as the stationary distribution of the position of a random walker who starts the walk in node v and then at each node either selects one of the outgoing connections or travels to the starting location. The probability (denoted p) of continuing the walk is a parameter of the personalized PageRank algorithm and is usually set to 0.85. Once calculated, the resulting PageRank vectors are normalized according to the Euclidean norm. The vectors are used to classify the nodes from which they were calculated.

For a single homogeneous network, the propositionalization results in one feature vector per node. For classifying a heterogeneous network, decomposed into k homogeneous networks, Grčar et al. [6] propose to concatenate and assign weights to the k vectors, obtained from the k homogeneous networks. The weights are optimized using a computationally expensive differential evolution [19]. A simpler alternative is to use equal weights and postpone weighting to the learning phase; due to the size of feature vectors in our experiments, we decided to follow this approach. Many classifiers, for example SVM classifier [4,14,16], kNN classifier [22] or a centroid classifier [7] can be used.

4   Methodology Improvement

We present two improvements to the methodology described in Sect. 3. First, we describe handling of imbalanced data sets, then we present a novel edge weighting approach used in the construction of homogeneous networks from the original network.

4.1   Imbalanced Data Sets and Label Propagation

The label propagation algorithm, as defined in [25], works by simply propagating class labels from each member, belonging to a certain class. By doing so, it seems possible that the algorithm may have a tendency to over-estimate the importance of larger classes (those with more instances) in the case when data is imbalanced. For example, Fig. 2 shows an example in which the label propagation algorithm will classify the central node as belonging to the first (larger) class, simply because it has three neighbors of class 1 and only two neighbors of class 2. This may, in some cases, not be the ideal outcome. It could be argued that the node we wish to classify is actually adjacent to *all* elements of class 1, but only *some* elements of class 2. Therefore, in this relative sense, class 1 nodes cast a stronger vote in favor of class 1 than nodes of class 2.

Using the reasoning described above, we see that there is a reason to believe that the label propagation approach may not perform well if the data is highly imbalanced, i.e., if the frequency of class labels are not approximately equal.

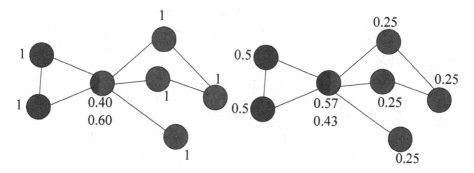

Fig. 2. Results of a label propagation algorithm on an imbalanced data set. If we run the label propagation algorithm as originally defined, each labeled node begin their iteration with a weight of 1 for the class they belong to. In each step of the iteration, every node collects the votes from its neighboring nodes and adds a portion (defined by α which was set to 0.6 in this example) of its original weight. We calculate that in this case, the central node receives a proportional vote of 0.40 from class 1 and a vote of 0.60 from class 2. However, using our modified weights, the labeled nodes start their iteration with a weight of $\frac{1}{2}$ for the class with 2 nodes and $\frac{1}{4}$ for the class with 4 nodes. Because of this, the proportion of votes for class 1 increases to 0.57. This is justified by the fact that the central node actually receives the highest possible vote that it can from a class consisting of only two nodes.

We propose an adjustment of the label propagation algorithm i.e., to change the initial label matrix Y so that larger classes have less effect in the iterative process. The value of the label matrix Y in this case is no longer binary, but it is set to $\frac{1}{|c_j|}$ if node i belongs to class j and 0 otherwise.

If the data set is balanced (all class values are equally represented), then the matrix Y is equivalent to the original binary matrix multiplied by the inverse of the class value size. This, along with (1), means that the resulting prediction matrix only changes by a constant and the final predictions remain unchanged. However, if the data set is imbalanced, smaller classes have a larger effect in the iterative calculation of F^*. This prevents votes of more frequent classes to outweigh votes of less frequent classes.

4.2   Text Mining Inspired Weights Calculation

We shortly present weighting of terms in the construction of bag-of-words (BOW) vectors and explain how the same ideas can be applied to extraction of homogeneous networks from heterogeneous networks.

Term Weighting in Text Mining. In bag-of-words vector construction one feature vector represents each document in a corpus of documents. In that vector, the i-th value corresponds to the i-th term (a word or a n-gram) that appears in the corpus. The value of the feature depends primarily on the frequency of

Table 1. Term weighing in text mining.

Scheme	Formula
tf	$f(t,d)$
if-idf	$f(t,d) \cdot \log \left(\dfrac{|D|}{|\{d' \in D : t \in d'\}|} \right)$
chi^2	$f(t,d) \cdot \sum_{c \in C} \left(\dfrac{(P(t \wedge c)P(\neg t \wedge \neg c) - P(t \wedge \neg c)P(\neg t \wedge c))^2}{P(t)P(\neg t)P(c)P(\neg c)} \right)$
ig	$f(t,d) \cdot \sum_{c \in C} \left(\sum_{c' \in \{c, \neg c\}} \left(\sum_{t' \in \{t, \neg t\}} \left(P(t', c') \cdot \log \dfrac{P(t' \wedge c')}{P(t')P(c')} \right) \right) \right)$
delta-idf	$f(t,d) \cdot \sum_{c \in C} \left(\log \dfrac{|c|}{|\{d' \in D : d' \in c \wedge t \in d'\}|} - \log \dfrac{|\neg c|}{|\{d' \in D : d' \notin c \wedge t \notin d'\}|} \right)$
rf	$f(t,d) \cdot \sum_{c \in C} \left(\log \left(2 + \dfrac{|\{d' \in D : d' \in c \wedge t \in d'\}|}{|\{d' \in D : d' \notin c \wedge t \notin d'\}|} \right) \right)$

the term in the particular document. We describe several methods for assigning the feature values. We use the following notations: $f(t,d)$ denotes the number of times a term t appears in the document d and D denotes the corpus (a set of documents). We assume that the documents in the set are labeled, each document belonging to a class c from the set of all classes C. We use the notation $t \in d$ to describe that a term t appears in document d. Where used, the term $P(t)$ is the probability that a randomly selected document contains the term t, and $P(c)$ is the probability that a randomly selected document belongs to class c. Table 1 shows different methods for term weighting. The term frequency (tf) weights each term with its frequency in the document. The term frequency–inverse document frequency (tf-idf) [10] addresses the drawback of the tf scheme, which tends to assign high values to common words that appear frequently in the corpus. The χ^2 (chi^2) weighting scheme [3] attempts to correct another drawback of the tf scheme (one which is not addressed by the tf-idf scheme) by taking also class value of processed documents into consideration. This allows the scheme to penalize terms that appear in documents of all classes, and favor terms which are specific to some classes. Information gain (ig) [3] uses class labels to improve term weights. The Δ-idf (delta-idf) [17] and the Relevance frequency (rf) [15] attempt to merge the ideas of idf and both above class-based schemes by penalizing both common and non-informative terms.

Midpoint Weighting in Homogeneous Network Construction. Let us revisit the example from Sect. 3.1, in which two papers are connected by one link for each author they share. The resulting network is equivalent to a network in which two papers are connected by a link with a weight equal to the number of authors that wrote both papers (Fig. 1). The method treats all authors equally which may not be correct. For example, if two papers share an author that only co-authored a small number of papers, it is more likely that these two papers are similar than if the two papers share an author that co-authored tens or even hundreds of papers. The first pair of papers should therefore be connected by a stronger weight than the second. Moreover, if the papers are labeled by the research field, then two papers, sharing an author publishing in only one research field, are more likely to be similar as if they share an author who has co-authored papers in several research fields. Again, the first pair of papers should be connected by the edge with larger weight.

Both described considerations are similar to the issues addressed in the term weighting schemes in document retrieval (presented at the beginning of this section). For example, the `tf-idf` weighting scheme attempts to decrease the weight of terms which appear in many documents, while we wish to decrease the weight of links, induced by authors which are connected to many papers. The `ig` weighting scheme decreases the weight of terms which appear in variously labeled documents, while we wish to decrease the weight of links induced by authors appearing in different research areas i.e., connected to variously labeled papers.

We alter the term weighting schemes in such a way that they can be used to set weights to midpoints in heterogeneous graphs (such as authors in our example). We propose that the weight of a link between two base nodes in the first step of the methodology (see Sect. 3) is calculated as the sum of weights of all the midpoints they share. In particular, if we construct a homogeneous network in which nodes are connected if they share a connection to a node of type T in the original heterogeneous network, then the weight of the link between nodes v and w should be equal to

$$\sum_{m \in T:(m,v) \in E \wedge (m,w) \in E} w(m),\qquad(2)$$

where $w(m)$ is the weight assigned to the midpoint m. The value of $w(m)$ can be calculated in several ways. Table 2 shows the proposed midpoint weighting heuristics corresponding to term weighting used in document retrieval (Table 1). The notation used was as follows. We denote with B the set of all nodes of the base node type, and with E the set of all edges of the heterogeneous network. When m is a node, $P(m)$ denotes the probability that a random base node is connected to the midpoint node m. We assume that nodes are labeled, each belonging to a class c from the set of all classes C. We use $P(c)$ to denote the probability that a random base node is in class c. The term $P(c \wedge m)$ denotes the probability that a random base node is both in class c and linked to midpoint m.

The `tf` weight is effectively used in [6], where all authors are weighed equally. The `delta-idf` weighting scheme, unlike other term weighting schemes, may assign negative weights to certain terms. Since link weights in graphs are assumed to be positive both by the PageRank and the link propagation algorithm, we must change the weighting scheme before it can be used to construct homogeneous networks. We propose that in the original weighting scheme, terms which receive negative values are deemed informative, as they are informative about the term *not* being typical of a certain class. Therefore it is reasonable to take the absolute values of the weights in network construction.

Table 2. Heuristics for weighting midpoints in homogeneous network construction.

Scheme	Formula
tf	1
if-idf	$\log\left(\dfrac{\|B\|}{\|\{b \in B : (b,m) \in E\}\|}\right)$
chi^2	$\displaystyle\sum_{c \in C} \dfrac{(P(m \wedge c)P(\neg m \wedge \neg c) - P(m, \neg c)P(\neg m, c))^2}{P(m)P(c)P(\neg m)P(\neg c)}$
ig	$\displaystyle\sum_{c \in C}\left(\sum_{c' \in \{c, \neg c\}}\left(\sum_{m' \in \{m, \neg m\}} P(m' \wedge c') \log\left(\dfrac{P(m' \wedge c')}{P(c')P(m')}\right)\right)\right)$
delta-idf	$\displaystyle\sum_{c \in C}\left\|\log\dfrac{\|c\|}{\|\{b' \in B : b' \in c \wedge (b',m) \in E\}\|} - \log\dfrac{\|\neg c\|}{\|\{b' \in B : b' \notin c \wedge (b',m) \notin E\}\|}\right\|.$
rf	$\displaystyle\sum_{c \in C}\left(\log\left(2 + \dfrac{\|\{b' \in B : b' \in c \wedge (b',m) \in E\}\|}{\|\{b' \in B : b' \notin c \wedge (b',m) \notin E\}\|}\right)\right)$

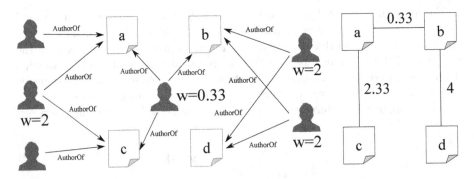

Fig. 3. The decomposition of the toy network from Fig. 1 using the χ^2 heuristic. The blue color denotes that the paper belongs to class 1 and the red color denotes class 2. (Color figure online)

Example 1. Figure 3 shows the decomposition of the network, seen in Fig. 1, using the χ^2 heuristic. The weight of the central author m is calculated as the sum over both classes of

$$\frac{(P(m \wedge c)P(\neg m \wedge \neg c) - P(m, \neg c)P(\neg m, c))^2}{P(m)P(c)P(\neg m)P(\neg c)} \tag{3}$$

When c is the first (blue) class, we can calculate the required values as $P(m \wedge c) = \frac{2}{4}, P(\neg m \wedge \neg c) = \frac{1}{4}, P(m, \neg c) = \frac{1}{4}, P(\neg m \wedge c) = 0; P(m) = \frac{3}{4}, P(\neg m) = \frac{1}{4}, P(c) = P(\neg c) = \frac{1}{2}$, yielding the first summand of 3 as

$$\frac{(\frac{2}{4} \cdot \frac{1}{4} - \frac{1}{4} \cdot 0)^2}{\frac{3}{4} \cdot \frac{1}{4} \cdot \frac{1}{2} \cdot \frac{1}{2}} = \frac{1}{3}.$$

When c is the second class, after calculating $P(m \wedge c) = P(\neg m \wedge \neg c) = P(m, \neg c) = P(\neg m \wedge c) = \frac{1}{4}$, we see that the second summand of 3 is 0 and the total weight of author m is $\frac{1}{3}$.

The weights of the remaining authors are calculated in the same way. In our case, none of the other authors wrote papers from both classes, so their weights are all equal to 2.

The decomposed network on the left is constructed by summing the weights of all common authors for each pair of papers. We see that the connection between papers a and b is now weaker because their common author was assigned a smaller weight.

5 Experimental Setting and Results

In this section, we describe the experiments used to evaluate the performance of the presented classifiers. We first describe the data set we used in our experiments. Then, we present the experiment set up and results.

5.1 Data Set Description

We evaluated the proposed weighting heuristics on a data set of customer purchases used in the PAKDD 2015 mining competition *Gender prediction based on e-commerce data*. The data consists of 30,000 customers. The data for each customer consists of the gender (the target variable), the start and end time of the purchase, and the list of products purchased. A typical product is described by a 4-part string (for example: A3/B5/C2/D8). The strings describe a 4-level hierarchy of products, meaning that the example product is the product $D8$ (or D-level category) which belongs to (A-level) category $A3$, sub-category (or B-level category) $B5$ and sub-subcategory (or C-level category) $C3$. The category levels are consistent, meaning that if two products belong to the same B-level category, they also belong to the same A-level category. The data set is highly imbalanced: 23,375 customers are women and 6,625 are men.

For the purpose of our experiments, we ignored the temporal aspects of the data set and only focused on the products the customers purchased. This allowed us to view the data set as an implicitly defined heterogeneous network. The network consists of five node types: customers (the base node type) and four hierarchy levels. In this heterogeneous network, every purchase by a customer defines four edges in the heterogeneous network: one edge between the customer and each (sub)category to which the product belongs.

We constructed four homogeneous networks from the original heterogeneous network. In the first, two customers are connected if they purchased the same product (same D-level item), i.e. if they are connected by a path in the original network that goes through a D-level item. In the second, they are connected if they purchased a product in the same sub-subcategory (C-level item), in the third if they purchased the same B-level item and in the fourth if they purchased the same A-level item. The constructed networks are referred to as A-, B-, C- and D-level networks from this point on.

5.2   Experiment Description

The first set of experiments was designed to determine if the results of [6], which show that a centroid classifier, trained on Personalized PageRank feature vectors, performs as good as the more complex SVM classifier. We tested the performance of the centroid classifier, the k-nearest neighbors classifier (with k set to $1, 2, 5$ and 10), and the SVM classifier. Because the data set is imbalanced we tested the SVM classifier both with uniform instance weights as well as weights proportional to the class frequencies. The tests were performed on feature vectors extracted from all four homogeneous networks. We randomly sampled 3,000 network nodes to train all classifiers and tested their performance on the remaining 27,000 nodes. The small size of the training set ensured the training phase was fast.

In the second set of experiments we tested the heuristics, used in the construction of the homogeneous networks. We tested three classifiers. The first was the SVM classifier using solely the Personalized PageRank vectors extracted from the network. As the results of the first experiment showed that weights, proportional to the class frequencies, improve the classification accuracy of the SVM classifier, we used the same weights for this set of experiments. The second classifier we tested was the label propagation classifier as defined in [25], which classified the network nodes using the graph itself. The third classifier was the label propagation classifier with the starting matrix Y, adjusted for the class frequencies, as proposed in Sect. 3.2. The goal of this round of experiments was to compare the label propagation classifier with the SVM classifier and evaluate whether the adjusted starting matrix Y has an effect on classifier performance. As in experiment 1, we trained the classifiers on a randomly sampled set of 3,000 nodes and tested their performance on the remaining 27,000 nodes.

In the third round of experiments, we tested the performance of the label propagation and propositionalization based classifiers on all four homogeneous networks. Based on the results of the first two sets of experiments, we used the SVM classifier for the propositionalization approach and the label propagation method with the modified starting matrix Y. As explained in Sect. 3.2, we constructed feature vectors for SVM classifiers by concatenating feature vectors of individual homogeneous networks. We constructed the adjacency matrix for the label propagation algorithm by summing the four adjacency matrices. One of the goals of the third round of experiments was to test the performance of the classifiers when they are trained on a large data set. This motivated us to train the classifiers on 90 % of the data set and test their performance on the remaining 10 %.

In all experiments we evaluated the accuracy of the classifiers using the *balanced accuracy* metric. This is the metric used in the PAKDD'15 Data Mining Competition and is defined as

$$\frac{\frac{|\{\text{Correctly classified male customers}\}|}{|\{\text{All male customers}\}|} + \frac{|\{\text{Correctly classified female customers}\}|}{|\{\text{All female customers}\}|}}{2}. \tag{4}$$

Table 3. Results of the three sets of experiments showing accuracies of the tested algorithms

Classifier:	Centroid	1-nn	2-nn	5-nn	10-nn	SVM	SVM (balanced weights)
A-level network	74.19%	63.61%	71.93	72.74%	74.36%	74.03%	74.62%
B-level network	70.78%	56.42%	59.17%	65.30%	67.73%	63.51%	72.61%
C-level network	64.71%	63.62%	67.21%	68.26%	71.65%	70.15%	75.18%
D-level network	60.08%	67.36%	70.39%	66.72%	66.06%	65.61%	71.17%

(a) Results of the first set of experiments.

Scheme	A-level	B-level	C-level	D-level
tf	76.61%	74.00%	77.34%	73.65%
chi^2	77.80%	74.17%	76.86%	68.76%
idf	77.80%	74.22%	77.23%	72.25%
delta	77.80%	74.14%	77.23%	72.52%
rf	77.80%	74.11%	76.81%	70.54%
ig	77.80%	74.12%	76.87%	68.72%

(b) Performance of the SVM classifier in the second round of experiments.

Scheme	A-level	B-level	C-level	D-level
tf	75.52%	64.28%	63.60%	72.44%
chi^2	76.02%	65.15%	71.95%	72.75%
idf	74.90%	63.83%	61.02%	72.48%
delta	74.90%	63.76%	61.05%	72.48%
rf	75.52%	64.28%	67.59%	72.55%
ig	76.02%	65.15%	72.41%	72.96%

(c) Performance of the label propagation classifier in the second round of experiments.

Scheme	A-level	B-level	C-level	D-level
tf	77.16%	74.75%	77.28%	73.91%
chi^2	77.16%	74.44%	77.61%	73.82%
idf	77.20%	74.70%	77.74%	73.76%
delta	77.20%	74.71%	77.74%	73.76%
rf	77.16%	74.59%	77.21%	74.03%
ig	77.16%	74.49%	77.59%	73.79%

(d) Performance of the balanced label propagation classifier in the second round of experiments.

Scheme	SVM	Label propagation
tf	81.35%	77.06%
chi^2	81.78%	77.10%
idf	82.09%	79.03%
delta	81.94%	79.08%
rf	81.49%	77.16%
ig	81.56%	77.12%

(e) The results of the third set of experiments showing the balanced accuracies of the SVM and label propagation classifiers on the entire data set.

5.3   Experimental Results

The first set of experiments, shown in Table 3a, shows that there is a large difference in the performance of different classifiers. Similarly to Grčar et al. [6], the simple centroid classifier performs well on feature vectors extracted from several different homogeneous networks. However, the classifier is still consistently outperformed by the SVM classifier if the instance weights of the classifiers are set according to the class sizes. We conclude that the optimal classifier for the methodology, introduced in [6], depends on the data set.

The results of the second set of experiments are shown in Table 3b, c and d. When comparing the results of the two label propagation approaches the results show that label propagation with adjusted starting matrix has large impact on the performance of the classifier, as the balanced accuracy increases by 1–2 % in the case of the A- and D-level network and even more in the case of B- and

C-level networks. This result confirms the intuition that, in Sect. 3.2, motivated the construction of the adjusted starting matrix.

Different heuristics used in construction of homogeneous networks also affect the final performance of all three classifiers. No heuristic consistently outperform the others, meaning that the choice of heuristic is application dependent. The last conclusion of the second round of experiments is that the computationally demanding propositionalization method does not outperform the label propagation method. In all four networks choosing the correct heuristic and correct weights for the starting matrix allows the label propagation method to perform comparably to the SVM classifier.

Table 3e shows the results of the third set of experiments. In this experiment, the propositionalization-based approach clearly outperforms the label propagation algorithm. It is possible that this effect occurs because the network propositionalization approach, in particular the SVM classifier, require more training examples (compared to the network propagation classifier) to perform well. A second explanation may come from the way the four networks were combined in our experiments (i.e. the concatenation approach in the case of network propositionalization and the matrix sum in the case of label propagation). By summing the adjacency matrices before performing label propagation, we implicitly assumed that the connections between customers that purchased the same D-level product, are equally important as connections between customers that purchased the same A-level product. This may cause the amount of A-level edges to everwhelm the effects of the D level edges, causing the resulting network to be very similar to the original A level network. The propositionalization based approach is less prone to an error of this type, as the idea of the SVM algorithm is to define correct weights for elements of the feature vectors. The SVM algorithm is therefore flexible enough to assign larger weights to the features, produced by the D-level network, if it estimates that these features are more important in classification. The second conclusion we can draw from the third set of experiments is that the effect of using weighting heuristics in the construction of the homogeneous networks is still obvious. With both classification methods the adjusted `delta` and `idf` heuristics perform best.

6   Conclusions and Further Work

While network analysis is a well established research field, analysis of heterogeneous networks is much newer and less researched. Methods taking the heterogeneous nature of the networks into account show an improved performance [2]. Some methods like RankClus and others presented in [20] are capable of solving tasks that cannot be defined with homogeneous information networks (like clustering two disjoint sets of entities). Another important novelty is combining network analysis with the analysis of node data, either in the form of text documents or results obtained from various experiments [5,6,8].

The contributions of the paper are as follows. By setting the weights of the initial class matrix proportionally to the class value frequency, we improved the performance of the label propagation algorithm when applied to a highly imbalanced data set. We adapted heuristics, developed primarily for use in text mining, for the construction of homogeneous networks from heterogeneous networks. Our results show that the choice of heuristics impacts the performance of both label propagation classifier and classifiers based on the propositionalization approach of [6]. We also present a variation of the label propagation approach, described in [25].

In future work, in-depth analysis of the network construction heuristics and their performance in classifiers applied to homogeneous networks will be pursued. We plan to design efficient methods for propositionalization of large data sets and decrease the computational load of PageRank calculations by first detecting communities in a network. Such "pre-processing" should reduce the size of a network on which PageRank calculations are to be performed.

We plan to test the methods, described in this paper, on publicly available data sets such as the DBLP, Cora and CiteSeer databases. The presented heuristics shall be evaluated on the methodology for mining text enriched heterogeneous networks presented in [6]. For that, one has to construct a heterogeneous network in which the central node represents genes, connected to the response of plants against an infection. We will enrich the nodes with papers from the PubMed database which mention the genes.

Acknowledgement. The presented work was partially supported by the European Commission through the Human Brain Project (Grant number 604102). We also acknowledge the support of research projects funded by the National Research Agency: the Knowledge Technologies research programme and the project Development and applications of new semantic data mining methods in life sciences.

References

1. Burt, R., Minor, M.: Applied Network Analysis: A Methodological Introduction. Sage Publications, Beverly Hills (1983)
2. Davis, D., Lichtenwalter, R., Chawla, N.V.: Multi-relational link prediction in heterogeneous information networks. In: Proceedings of the 2011 International Conference on Advances in Social Networks Analysis and Mining, pp. 281–288 (2011)
3. Debole, F., Sebastiani, F.: Supervised term weighting for automated text categorization. In: Sirmakessis, S. (ed.) Text Mining and Its Applications, pp. 81–97. Springer, Heidelberg (2004)
4. D'Orazio, V., Landis, S.T., Palmer, G., Schrodt, P.: Separating the wheat from the chaff: applications of automated document classification using support vector machines. Polytical Anal. **22**(2), 224–242 (2014)
5. Dutkowski, J., Ideker, T.: Protein networks as logic functions in development and cancer. PLoS Comput. Biol. **7**(9), e1002180 (2011)
6. Grčar, M., Trdin, N., Lavrač, N.: A methodology for mining document-enriched heterogeneous information networks. Comput. J. **56**(3), 321–335 (2013)
7. Han, E.-H., Karypis, G.: Analysis and experimental results. In Proceedings of the 4th European Conference on Principles of Data Mining and Knowledge Discovery, pp. 424–431. Springer (2000)

8. Hofree, M., Shen, J.P., Carter, H., Gross, A., Ideker, T.: Network-based stratification of tumor mutations. Nat. Methods **10**(11), 1108–1115 (2013)
9. Jeh, G., Widom, J.: A measure of structural-context similarity. In: Proceedings of the 8th ACM SIGKDD International Conference on Knowledge Discovery and Data Mining, pp. 538–543. ACM (2002)
10. Jones, K.S.: A statistical interpretation of term specificity and its application in retrieval. J. Doc. **28**, 11–21 (1972)
11. Kleinberg, J.M.: Authoritative sources in a hyperlinked environment. J. ACM **46**(5), 604–632 (1999)
12. Kondor, R.I., Lafferty, J.D.: Diffusion kernels on graphs and other discrete input spaces. In: Proceedings of the 19th International Conference on Machine Learning, pp. 315–322 (2002)
13. Kralj, J.: Mining heterogeneous citation networks. In: Proceedings of the 19th Pacific-Asia Conference on Knowledge Discovery and Data Mining, pp. 672–683 (2015)
14. Kwok, J.T.-Y.: Automated text categorization using support vector machine. In: Proceedings of the 5th International Conference on Neural Information Processing, pp. 347–351 (1998)
15. Lan, M., Tan, C.L., Su, J., Lu, Y.: Supervised and traditional term weighting methods for automatic text categorization. IEEE Trans. Pattern Anal. Mach. Intell. **31**(4), 721–735 (2009)
16. Manevitz, L.M., Yousef, M.: One-class SVMs for document classification. J. Mach. Learn. Res. **2**, 139–154 (2002)
17. Martineau, J., Finin, T.: An improved feature space for sentiment analysis. In: Proceedings of the Third AAAI Internatonal Conference on Weblogs and Social Media, San Jose, CA. AAAI Press (2009)
18. Page, L., Brin, S., Motwani, R., Winograd, T.: The PageRank citation ranking: bringing order to the web. Technical report, Stanford InfoLab (1999)
19. Storn, R., Price, K.: Differential evolution: a simple and efficient heuristic for global optimization over continuous spaces. J. Global Optim. **11**(4), 341–359 (1997)
20. Sun, Y., Han, J.: Principles and Methodologies. Morgan & Claypool Publishers, San Rafael (2012)
21. Sun, Y., Yu, Y., Han, J.: Ranking-based clustering of heterogeneous information networks with star network schema. In: Proceedings of the 15th ACM SIGKDD International Conference on Knowledge Discovery and Data Mining, pp. 797–806 (2009)
22. Tan, S.: An effective refinement strategy for KNN text classifier. Expert Syst. Appl. **30**(2), 290–298 (2006)
23. Vanunu, O., Magger, O., Ruppin, E., Shlomi, T., Sharan, R.: Associating genes and protein complexes with disease via network propagation. PLoS Comput. Biol. **6**(1), e1000641 (2010)
24. Zachary, W.: An information flow model for conflict and fission in small groups. J. Anthropol. Res. **33**, 452–473 (1977)
25. Zhou, D., Bousquet, O., Lal, T.N., Weston, J., Schölkopf, B.: Learning with local and global consistency. Adv. Neural Inf. Process. Syst. **16**(16), 321–328 (2004)

Sequences

Semi-supervised Multivariate Sequential Pattern Mining

Zhao Xu[1]([✉]), Koichi Funaya[1], Haifeng Chen[2], and Sergio Leoni[1]

[1] NEC Laboratories Europe, Kurfürsten-Anlage 36, 69115 Heidelberg, Germany
{zhao.xu,koichi.funaya,sergio.leoni}@neclab.eu
[2] NEC Laboratories America, 4 Independence Way, Suite 200,
Princeton, NJ 08540, USA
haifeng@nec-labs.com

Abstract. Multivariate sequence analysis is of growing interest for learning on data with numerous correlated time-stamped sequences. It is characterized by correlations among dimensions of multivariate sequences and may not be separately analyzed as multiple independent univariate sequences. On the other hand, labeled data is usually expensive and difficult to obtain in many real-world applications. We present a graph-based semi-supervised learning framework for multivariate sequence classification. The framework explores the correlation within the multivariate sequences, and exploits additional information about the distribution of both labeled and unlabeled data to provide better predictive performance. We also develop an efficient method to extend the graph-based learning approach to out-of-sample prediction. We demonstrate the effectiveness of our approach on real-world multivariate sequence datasets from three domains.

1   Introduction

The growing popularity of social media, e-commerce and sensor systems has generated considerable interest in sequential data analysis, ranging from web site visit data mining to sensor data analysis in Internet of Things systems. The sequential data in many applications involves numerous correlated observations received at each time point, such as data collected from multiple related sensors. An intrinsic property of multivariate sequential data is the correlations between different dimensions. So it may not be modeled as multiple independent univariate sequences.

In many real-world applications, there is usually a large amount of unlabeled data but limited labeled data, which can be difficult and time consuming to obtain. Most existing work on semi-supervised sequence classification has focused on univariate sequential data, such as Wei et al.'s work based on self-training [26], and the SUCCESS method based on clustering [15]. We present a graph-based semi-supervised learning framework for multivariate sequence classification. The framework explores the correlation across dimensions of multivariate sequences, and exploits the underlying structure of both labeled and unlabeled sequential

© Springer International Publishing Switzerland 2016
M. Ceci et al. (Eds.): NFMCP 2015, LNAI 9607, pp. 211–223, 2016.
DOI: 10.1007/978-3-319-39315-5_14

data. In our framework, a weighted graph is constructed based on similarity of both labeled and unlabeled sequences to capture the underlying structures of the whole dataset. To effectively encode the correlations between different dimensions of multivariate sequences, we employ a PCA (principal component analysis) based similarity measure. The unlabeled sequences are classified by label propagation over the constructed graph with a harmonic Gaussian field based method [31]. In addition, we develop an effective method to extend the graph-based learning method to out-of-sample prediction. We evaluate the framework with real-world multivariate sequence datasets from three domains. The results demonstrate superior predictive performance of the proposed approach.

The rest of the paper is organized as follows. We discuss related work in Sect. 2. Section 3 introduces the proposed semi-supervised multivariate sequence classification framework. The experimental results are provided in Sect. 4. Finally the paper is concluded in Sect. 5.

2   Related Work

Semi-supervised algorithms that learn from both labeled and unlabeled data have attracted much interest and various approaches have been proposed [4,19,30], such as self-training [18], co-training [3], and graph-based methods [31]. Some semi-supervised learning methods for sequential data classification have been proposed in the recent literature to improve the predictive performance by leveraging unlabeled data. Wei et al. introduced a self-training method based on one-nearest-neighbor classifier [26]. The method starts by training a classifier with labeled data, by which the unlabeled data is classified. Then the most confident unlabeled sequences with the estimated labels are added to the training data. The classifier is retrained and the procedure repeated. Self-training based methods are dependent on the employed classifier, and the classification mistake can reinforce itself [4]. The SUCCESS method proposed by Marussy et al. is based on constrained hierarchical clustering [15]. The method clusters the whole set of sequences, including labeled and unlabeled ones, using single-linkage hierarchical agglomerative clustering. Then the top-level clusters are labeled by their corresponding seeds. Clustering based semi-supervised learning methods usually rely on whether clustering algorithms can match the true data distributions [30]. These work on semi-supervised sequence classification mainly concentrate on univariate sequential data, whereas multivariate sequential data is of growing importance in many applications.

3   The Semi-supervised Learning Framework

In this section, we will describe the graph-based semi-supervised learning framework for multivariate sequence classification. Assume that there are a set of L labeled sequences, denoted as $\{(ts_1, y_1), (ts_2, y_2), \ldots, (ts_L, y_L)\}$, and a set of U unlabeled sequences, denoted as $\{ts_{L+1}, ts_{L+2}, \ldots, ts_{L+U}\}$. The sequences are evenly spaced, and can be of variable length. The length of a sequence ts_i is denoted as T_i. Each sequence is K-dimensional, and can be represented as a matrix of size $T_i \times K$. The goal of the proposed framework is to fit a model to predict unknown labels $\{y_{L+1}, y_{L+2}, \ldots, y_{L+U}\}$ by exploiting underlying structure of the entire data explored with a weighted graph $\mathcal{G} = (V, E, W)$. Each multivariate sequence in the data is denoted as a vertex $v \in V$ of the graph $\mathcal{G}$, and all the sequences are linked with each other using undirected edges $e \in E$, which are weighted by a function of distance between the involved sequences. The unknown labels can be estimated by label propagation over the graph. In the following subsections, we will provide more details about graph construction, label propagation, and an out-of-sample extension.

3.1   Graph Construction

To effectively discover and exploit the underlying structures of a set of multivariate sequences, we construct a weighted graph $\mathcal{G}$ represented as an adjacency matrix W of size $(L+U) \times (L+U)$, where L and U denote the numbers of labeled and unlabeled sequences, respectively. Each entry $W_{i,j}$ represents the weight of an undirected edge $e_{i,j}$ between the sequences ts_i and ts_j. Intuitively, the larger the weight, the smaller the distance/dissimilarity between the two sequences, and thus the more likely the sequences have the same labels. We formulate the weight as a function of the corresponding distance $d_{i,j}$, i.e.

$$W_{i,j} = f(d_{i,j}),\ d_{i,j} \geq 0.$$

The function $f(\cdot)$ should be non-negative and monotonically decreasing. The weight $W_{i,j}$ vanishes as $d_{i,j} \to \infty$. Inspired with stationary kernels, $W_{i,j}$ can be defined as, e.g.:

$$\text{Squared exponential:}\ \exp(-d^2/2\ell^2) \qquad (1)$$

$$\text{Rational quadratic:}\ (1 + d^2/2\alpha\ell^2)^{-\alpha} \qquad (2)$$

$$\gamma\text{-exponential:}\ \exp(-(d/\ell)^\gamma),\ 0 < \gamma \leq 2 \qquad (3)$$

The function $f(\cdot)$ quantifies the decay of the weight with increasing distance. Figure 1 shows two examples of weight functions. One can find that with the same parameter setting, squared exponential weight decreases slower than γ-exponential one with $\gamma = 1$ when the distance is small, but faster when the distance gets large.

The adjacency matrix W is constructed on the basis of the distance among multivariate sequences that can well capture the characteristics of sequences and the underlying structure of the dataset. The distance measure for univariate sequences has been largely investigated in the literature [7,14,16,25]. However these methods can not be simply extended to multivariate sequences due to the correlation between different dimensions of multivariate sequences. The multivariate sequence distance measures fall into two categories: dynamic time warping (DTW) based methods [1,17] and PCA/SVD based methods [10,13,20,27,29]. In this work, we exploit the Eros method [29], which explores

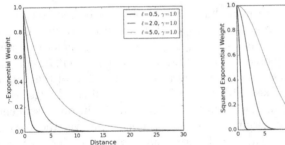
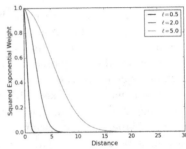

Fig. 1. Example weight functions: γ-exponential weight (left) and squared exponential weight (right).

correlations between different dimensions of multivariate sequences based on principal component analysis (PCA), and outperforms the distance measures based on DTW and PCA similarity factor [29]. The Eros method extends Frobenius matrix norm to measure similarity between two sequences. As described above, each multivariate sequence can be represented as a matrix of size $Ti \times K$. With PCA method, we can find a K number of principal components (eigenvectors) for each sequence. The similarity of two sequences ts_i and ts_j is computed as a weighted sum of cosine similarity of their corresponding principal components:

$$Eros(i,j) = \sum_{k=1}^{K} w_k |\cos \theta_k|, \tag{4}$$

where θ_k denotes the angle between the k'th principal components of ts_i and ts_j. The parameter w_k specifies how much variability of the set of sequences can be explained by the k'th principal components:

$$w_k = \frac{\sum_{i'}^{N} \lambda_{k,i'}}{\sum_{k}^{K} \sum_{i'}^{N} \lambda_{k,i'}}, \tag{5}$$

where $\lambda_{k,i'}$ is the eigenvalue associated with the k'th principal component of the sequence $ts_{i'}$. Based on the equation, one can find that the Eros similarity ranges between 0 and 1. As the Eros method measures the similarity between two multivariate sequences, we define the distance as $d_{i,j} = 1 - Eros(i,j)$ for graph construction. Alternatively, one can directly use similarity between sequences as the weight of the adjacency matrix W. It is somewhat equal to a linear kernel, which may not be as flexible as exponential kernels. For more details about kernel selection, please refer to [5,6,12].

In summary, the adjacency matrix W formulates the constructed graph for the entire multivariate sequence dataset. Each entry $W_{i,j}$ represents the weight of an edge $e_{i,j}$, and is modeled as a function of the distance between the involved sequences ts_i and ts_j. The larger the distance, the smaller the weight $W_{i,j}$. All the entries of W have to be non-negative and symmetric, but the matrix W itself is not necessarily positive semi-definite.

3.2   Label Propagation

After formulating graph representation $\mathcal{G} = (V, E, W)$ of a set of sequences, we can model the underlying structures of the data by exploring the adjacency matrix W. The unlabeled sequences can then be classified by propagating the label information from the labeled sequences to unknown ones over the constructed graph. In this work, we use a harmonic Gaussian field based method [31] for label propagation.

Each multivariate sequence is associated with an auxiliary random variable $z_i \in \mathbb{R}$, which represents soft label of the sequence. The distribution of the set of random variables z_i's is modeled as Gaussian random field. In particular, the state of z_i is only conditioned on the connected random variables, and follows a Gaussian distribution. Intuitively if two sequences are connected with a high weight edge, then they most likely have similar labels. To formulate the intuition, the energy, i.e. sum of clique potentials of a Markov random field, can be defined as [31]:

$$E(\mathbf{z}) = \frac{1}{4} \sum_{i,j} W_{i,j}(z_i - z_j)^2. \tag{6}$$

Therefore, the distribution of the Gaussian field is

$$p_\beta(\mathbf{z}) \propto \exp\left(-\beta E(\mathbf{z})\right),$$

$$= \exp\left(-\frac{\beta}{4} \sum_{i,j} W_{i,j}(z_i - z_j)^2\right)$$

$$= \exp\left(-\frac{\beta}{2} \mathbf{z}^T \Delta \mathbf{z}\right) \tag{7}$$

where the parameter β is a positive real number. Δ denotes combinatorial graph Laplacian, which is an essential component in graph-based methods:

$$\Delta = D - W, \tag{8}$$

where D is diagonal degree matrix with $D_{i,i} = \sum_j W_{i,j}$. The minimum energy of the field can be achieved at $\mathbf{z}^*$, which satisfies

$$\Delta \mathbf{z}^* = 0 \tag{9}$$

due to harmonic property of the field [31]. Intuitively, as part of $\mathbf{z}$ is known ($\mathbf{z}_\ell$ of labeled sequences), the function $\mathbf{z}^T \Delta \mathbf{z}$ is minimized when its derivative $\Delta \mathbf{z}$ is zero. To characterize the properties of the data explicitly in terms of matrix operations, the harmonic function (9) is expanded as:

$$\begin{bmatrix} D_{\ell,\ell} - W_{\ell,\ell} & -W_{\ell,u} \\ -W_{u,\ell} & D_{u,u} - W_{u,u} \end{bmatrix} \begin{bmatrix} \mathbf{z}_\ell^* \\ \mathbf{z}_u^* \end{bmatrix} = \mathbf{0}, \tag{10}$$

where $\mathbf{z}_\ell^*$ denotes the observed labels, and $\mathbf{z}_u^*$ is the unknown ones to be predicted. The Laplacian of the entire data is split into four corresponding blocks for labeled and unlabeled sequences. By solving the function (10), we have:

$$\mathbf{z}_u^* = (D_{u,u} - W_{u,u})^{-1} W_{u,\ell} \mathbf{z}_\ell^*. \tag{11}$$

It is eqivalent to the maximum likelihood estimation of $\mathbf{z}_u$ given the observed labels $\mathbf{z}_\ell$ and the adjacency matrix W. As the sequence labels $\mathbf{z}$ follow a Gaussian distribution (7), the conditional distribution $p(\mathbf{z}_u|\mathbf{z}_\ell)$ is still Gaussian, which mean (11) is the maximum likelihood estimation of $\mathbf{z}_u$. The energy of the Gaussian field is minimized at $\mathbf{z}_u = \mathbf{z}_u^*$. The class assignment of unlabeled sequences can then be obtained by thresholding the auxiliary variables $\mathbf{z}_u^*$.

Additionally, a graph kernel can be obtained using the Laplacian Δ. In particular, the distribution of the Gaussian random field is:

$$p_\beta(\mathbf{z}) \propto \exp\left(-\frac{\beta}{2}\mathbf{z}^T \Delta \mathbf{z}\right). \tag{12}$$

The term $(\beta\Delta)^{-1}$ specifies a graph kernel (covariance matrix). Considering the singular matrix issue, we further regularize the Laplacian as

$$\Delta = \Delta + \epsilon \mathbf{I},$$

where $\mathbf{I}$ denotes identity matrix and ϵ is a small positive real number. It corresponds to remove zero eigenvalues of the Laplacian by regularizing its spectrum [22]. Finally, the graph kernel is computed as:

$$k(\mathcal{G}) = (\beta\Delta + \epsilon \mathbf{I})^{-1}. \tag{13}$$

Due to the inverse operation in the kernel computation, the covariance between any two sequences is not a local one, but depends on the whole set of sequences. Intuitively, this kernel can better capture the underlying structure of a set of sequences.

3.3   Extension to Out-of-Samples

The graph Laplacian based label propagation method defines a label distribution over a fixed dataset, and does not provide a natural way to make prediction on labels of newly available data. The commonly used out-of-sample extensions [2,21,24] are generally based on kernel methods, and often involve complex computation. Here we introduce a fast approximate method to extend the graph based learning method from a transductive setting to an inductive one. The method is based on the probabilistic Gaussian field model and obtain a consistent out-of-sample solution for novel data.

Assume that there are a new set of O unlabeled sequences, denoted as $\{ts_{L+U+1}, ts_{L+U+2}, \ldots, ts_{L+U+O}\}$. We extend the Gaussian field defined with

(6) and (7) to include the out-of-samples. The harmonic function (10) is then updated for the newly available data:

$$\begin{bmatrix} D_{\ell,\ell} - W_{\ell,\ell} & -W_{\ell,u} & -W_{\ell,o} \\ -W_{u,\ell} & D_{u,u} - W_{u,u} & -W_{u,o} \\ -W_{o,\ell} & -W_{o,u} & D_{o,o} - W_{o,o} \end{bmatrix} \begin{bmatrix} \mathbf{z}_\ell^* \\ \mathbf{z}_u^* \\ \mathbf{z}_o^* \end{bmatrix} = \mathbf{0}, \tag{14}$$

where $\mathbf{z}_o^*$ denotes the unknown labels of the out-of-sample data. The Laplacian matrix is split into nine blocks for labeled, unlabeled, and out-of-sample sequences. Solving the harmonic function, we obtain:

$$\begin{bmatrix} \mathbf{z}_u^* \\ \mathbf{z}_o^* \end{bmatrix} = \begin{bmatrix} D_{u,u} - W_{u,u} & -W_{u,o} \\ -W_{o,u} & D_{o,o} - W_{o,o} \end{bmatrix}^{-1} \begin{bmatrix} W_{u,\ell} \\ W_{o,\ell} \end{bmatrix} \mathbf{z}_\ell^* \tag{15}$$

Compared with the solution (11) for the transductive setting, it seems that we have to compute a matrix inverse from scratch and can not exploit the transductive results to accelerate the computation. To address the challenge, we develop a fast approximate method by converting (15) into a linear equation system

$$\begin{bmatrix} D_{u,u} - W_{u,u} & -W_{u,o} \\ -W_{o,u} & D_{o,o} - W_{o,o} \end{bmatrix} \begin{bmatrix} \mathbf{z}_u^* \\ \mathbf{z}_o^* \end{bmatrix} = \begin{bmatrix} W_{u,\ell} \mathbf{z}_\ell^* \\ W_{o,\ell} \mathbf{z}_\ell^* \end{bmatrix}. \tag{16}$$

We now obtain two equations:

$$(D_{u,u} - W_{u,u})\mathbf{z}_u^* - W_{u,o}\mathbf{z}_o^* = W_{u,\ell}\mathbf{z}_\ell^* \tag{17}$$

$$(D_{o,o} - W_{o,o})\mathbf{z}_o^* - W_{o,u}\mathbf{z}_u^* = W_{o,\ell}\mathbf{z}_\ell^* \tag{18}$$

As we have computed $\mathbf{z}_u^*$ in the transductive setting with (11), the labels of out-of-samples can be approximated by solving (18):

$$\mathbf{z}_o^* = (D_{o,o} - W_{o,o})^{-1}(W_{o,\ell}\mathbf{z}_\ell^* + W_{o,u}\mathbf{z}_u^*). \tag{19}$$

The fast approximation method exploits the results of the transductive setting, and can largely reduce the computational cost for out-of-sample prediction from $\mathcal{O}((U+O)^3)$ to $\mathcal{O}(O^3)$. As an approximation method, the out-of-sample extension introduces additional residual $W_{u,o}\mathbf{z}_o^*$ to the linear equation system (comparing the Eqs. (11) and (17)). When the amount of the newly available data is small $O << L+U$, the residual is negligible and the approximation (19) is close to the exact solution. When the amount of out-of-sample data is relatively large and the distribution of the entire data deviates from the one estimated in the transductive setting, we can still exploit the results (11) and (19) by employing them as initial guess of an iterative method for the linear Eq. (16). For example, we can use Jacobi iteration [9]:

$$\mathbf{z}_{u,o}^* = A_1^{-1}(b - A_2 \mathbf{z}_{u,o}^*)$$

$$A_1 = diag(A), \quad A_2 = A - A_1$$

$$\mathbf{z}_{u,o}^* = \begin{bmatrix} \mathbf{z}_u \\ \mathbf{z}_o \end{bmatrix}, \quad b = \begin{bmatrix} W_{u,\ell}\mathbf{z}_\ell^* \\ W_{o,\ell}\mathbf{z}_\ell^* \end{bmatrix}, \quad A = \begin{bmatrix} D_{u,u} - W_{u,u} & -W_{u,o} \\ -W_{o,u} & D_{o,o} - W_{o,o} \end{bmatrix}. \tag{20}$$

Since A is a submatrix of the Laplacian of the entire data, we have $|a_{i,i}| > \sum_{j=1, j\neq i}^{U+O} |a_{i,j}|$. Thus the Jacobi iteration will converge to $\mathbf{z}_{u,o}^* = A^{-1}b$ [9]. The results of (11) and (19) provide an informative initialization and can thus largely reduce the learning iterations.

4   Experimental Analysis

We evaluate the proposed framework for multivariate sequence classification with three benchmark datasets, including: (1) network traffic data for network application identification, (2) position tracker data for sign language recognition, and (3) speech data for speaker identification. All the datasets used in the experiments are publicly available.

The network traffic data (NetFlow) contains network flow data which is defined as a series of network packets transferred between IP-Port pairs [23]. The length and the transfer direction of a packet are selected as features. The task of the experiments is to classify the network flows into two different applications: *Skype* and *BitTorrent*. Figure 2 visualizes some examples of network flows for different network applications.

The Australian sign language data (AusLan) consists of samples of 95 types of signs. These examples are captured from a native signer using high-quality position trackers with totally 22 sensors [8]. The signs are of variable length. Figure 3 illustrates some examples for different signs. The task of the experiments is to assign the 22-dimensional sequences into one of 95 signs (classes).

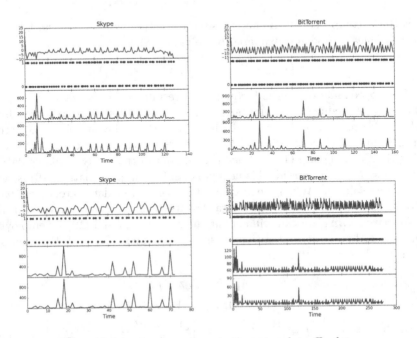

Fig. 2. Multivariate sequences in the network traffic data.

The Japanese vowel dataset (JapVow) includes nine speakers, who utter two Japanese vowels successively [11]. Each utterance by a speaker is recorded with a sequence of LPC cepstrum coefficients. The length of the sequences is variable, and each point of the sequences is of 12 features. The task of the experiments is to identify the nine speakers (classes) based on the 12-dimensional sequences.

The proposed framework is compared with the k nearest neighbor classifier (denoted as kNN with $k = 1, 3, 5$), which is a commonly-used method in sequence classification problem with superior predictive performance [28]. We also use the Eros distance for the kNN method in the experiments. For the proposed framework, the construction of sequence graphs is based on γ-exponential function with $\gamma = 1$. Every experiment is repeated 10 times, the averaged misclassification ratio is reported to measure the performance.

We first investigate the classification performance of the proposed framework for multivariate sequences in transductive setting. In the experiments, we randomly select some sequences as labeled data and the rest as unlabeled ones to predict. The experimental results on the three datasets are summarized in Table 1. The experiments show that our framework provides superior predictive performance in transductive setting, especially when the amount of the labeled training data is small. The experimental analysis demonstrates that exploiting the underlying structure detected from the entire data does improve the prediction performance.

We also perform experimental analysis on the proposed approximate inductive learning method, and evaluate how well the method can propagate label information to out-of-samples sequences. The experimental results on the three

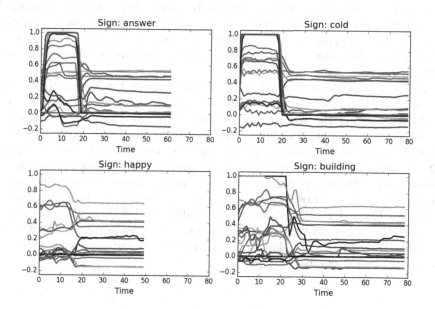

Fig. 3. Multivariate sequences in the AusLan data.

Table 1. Misclassification ratio for the transductive setting

	NetFlow				
Num. of labeled sequences	20	40	60	80	100
Our approach	0.201	0.18	0.178	0.168	0.163
kNN, $k = 1$	0.283	0.217	0.203	0.192	0.178
kNN, $k = 3$	0.244	0.211	0.184	0.172	0.173
kNN, $k = 5$	0.249	0.217	0.202	0.177	0.171
	AusLan				
Num. of labeled sequences	100	200	300	400	500
Our approach	0.320	0.236	0.197	0.171	0.167
kNN, $k = 1$	0.382	0.295	0.245	0.217	0.201
kNN, $k = 3$	0.394	0.325	0.263	0.228	0.210
kNN, $k = 5$	0.425	0.368	0.296	0.253	0.228
	JapVow				
Num. of labeled sequences	20	40	60	80	100
Our approach	0.343	0.302	0.290	0.270	0.271
kNN, $k = 1$	0.477	0.408	0.372	0.349	0.348
kNN, $k = 3$	0.480	0.406	0.372	0.343	0.337
kNN, $k = 5$	0.525	0.417	0.388	0.356	0.339

datasets are shown as Table 2. We randomly select some sequences as labeled ones, and then about 10 % of the rest sequences as out-of-samples. One can find that, by leveraging the underlying structure discovered in the transductive learning procedure, the proposed inductive method achieves superior prediction performance.

Our framework also demonstrates advantages when the amount of out-of-sample data becomes relatively large. We perform experimental analysis by randomly selecting 100 sequences as labeled data, and 10 % (20 % and 30 %) of the rest data as out-of-samples. With increasing number of out-of-samples, the distribution of the whole data more likely deviates from the one estimated with transductive learning. In this case, the proposed framework converts label propagation to finding optimal solution for a linear equation system. The transductive results are leveraged as initialization of an iterative method (e.g., Jacobi iteration). Figure 4 visualizes averaged log difference between labels predicted by iterative method and the ones re-computed by label propagation with the Eq. (15). The results show that the iterative method can effectively adapt to increasing amount of out-of-sample data.

Table 2. Misclassification ratio for the inductive setting

	NetFlow				
Num. of labeled sequences	20	40	60	80	100
Our approach	0.200	0.181	0.168	0.167	0.162
kNN, $k=1$	0.278	0.204	0.192	0.197	0.188
kNN, $k=3$	0.250	0.208	0.174	0.172	0.173
kNN, $k=5$	0.238	0.211	0.202	0.183	0.178
	AusLan				
Num. of labeled sequences	100	200	300	400	500
Our approach	0.320	0.250	0.210	0.178	0.174
kNN, $k=1$	0.382	0.293	0.245	0.215	0.200
kNN, $k=3$	0.396	0.336	0.277	0.235	0.219
kNN, $k=5$	0.430	0.371	0.301	0.261	0.232
	JapVow				
Num. of labeled sequences	20	40	60	80	100
Our approach	0.342	0.310	0.290	0.270	0.263
kNN, $k=1$	0.488	0.407	0.372	0.352	0.345
kNN, $k=3$	0.482	0.408	0.347	0.335	0.337
kNN, $k=5$	0.537	0.408	0.367	0.343	0.343

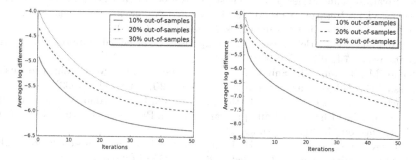

Fig. 4. Learning curve of Jacobi iterative method for inductive learning on the NetFlow (left) and JapVow (right) data.

5   Conclusion

In this paper we present a semi-supervised learning framework to improve the performance of multivariate sequence classification. The framework exploits the underlying structure of both labeled and unlabeled sequential data with graph-based approach. Meanwhile the correlation between different dimensions of multivariate sequences is also effectively incorporated in graph construction. In addition, an efficient and coherent method is developed to extend the graph-based learning to inductive settings. Our method can leverage the underlying structure discovered with transductive learning and adapt to increasing amount of out-of-sample data. The proposed method achieves promising predictive performance on multiple real-world multivariate sequential datasets. An interesting avenue for future work would be to extend the framework to other types of time stamped sequential data, such as unevenly spaced sequence data.

References

1. Banko, Z., Abonyi, J.: Correlation based dynamic time warping of multivariate time series. Expert Syst. Appl. **39**(17), 12814–12823 (2012)
2. Belkin, M., Niyogi, P., Sindhwani, V.: On manifold regularization. In: Proceedings of the 10th International Workshop on Artificial Intelligence and Statistics (2005)
3. Blum, A., Mitchell, T.: Combining labeled and unlabeled data with co-training. In: Proceedings of the Workshop on Computational Learning Theory (1998)
4. Chapelle, O., Schölkopf, B., Zien, A. (eds.): Semi-Supervised Learning. MIT Press, Cambridge (2006)
5. Chapelle, O., Vapnik, V.: Model selection for support vector machines. In: NIPS (1999)
6. Cristianini, N., Kandola, J., Elisseeff, A., ShaweTaylor, J.: On kernel target alignment. In: NIPS (2001)
7. Esling, P., Agon, C.: Time-series data mining. ACM Comput. Surv. **45**(1), 1–34 (2012)
8. Kadous, M.W.: Temporal Classification: Extending the Classification Paradigm to Multivariate Time Series. Ph.D. Thesis, University of New South Wales (2002)
9. Kelley, C.T.: Iterative Methods for Linear and Nonlinear Equations. Society for Industrial and Applied Mathematics, Philadelphia (1995)
10. Krzanowski, W.J.: Between-groups comparison of principal components. J. Am. Stat. Assoc. **74**, 703–707 (1979)
11. Kudo, M., Toyama, J., Shimbo, M.: Multidimensional curve classification using passing-through regions. Pattern Recogn. Lett. **20**, 1103–1111 (1999)
12. Lanckriet, G., Cristianini, N., Bartlett, P., Ghaoui, L.E., Jordan, M.: Learning the kernel matrix with semi-definite programming. In: Proceedings of the International Conference on Machine Learning (2002)
13. Li, C., Khan, L., Prabhakaran, B.: Real-time classification of variable length multi-attribute motions. Knowl. Inf. Syst. **10**(2), 16317183 (2005)
14. Liao, T.W.: Clustering of time series data-a survey. Pattern Recogn. **38**(11), 1857–1874 (2005)
15. Marussy, K., Buza, K.: SUCCESS: a new approach for semi-supervised classification of time-series. In: Rutkowski, L., Korytkowski, M., Scherer, R., Tadeusiewicz, R., Zadeh, L.A., Zurada, J.M. (eds.) ICAISC 2013, Part I. LNCS, vol. 7894, pp. 437–447. Springer, Heidelberg (2013)
16. Montero, P., Vilar, J.A.: Tsclust: An r package for time series clustering. J. Stat. Softw. **62**(1), 1–43 (2014)
17. Rath, T.M., Manmatha, R.: Lower-bounding of dynamic time warping distances formultivariate time series. Technical report MM-40, University of Massachusetts (2002)
18. Scudder, H.J.: Probability of error of some adaptive pattern-recognition machines. IEEE Trans. Inf. Theory **11**, 363–371 (1965)

19. Seeger, M.: Learning with labeled and unlabeled data. Technical report, Institute for ANC, Edinburgh, UK (2001)
20. Shahabi, C., Yan, D.: Real-time pattern isolation and recognition over immersive sensor data streams. In: Proceedings of the 9th International Conference on Multi-Media Modeling (2003)
21. Sindhwani, V., Niyogi, P., Belkin, M.: Beyond the point cloud: from transductive to semi-supervised learning. In: Proceedings of the 22nd International Conference on Machine Learning (2005)
22. Smola, A.J., Kondor, R.: Kernels and regularization on graphs. In: Proceedings of the Conference on Learning Theory (2003)
23. Subakan, Y.C., Kurt, B., Cemgil, A.T., Sankur, B.: Probabilistic sequence clustering with spectral learning. Digit. Sig. Proc. **29**, 1–19 (2014)
24. Wang, F., Zhang, C.: Label propagation through linear neighborhoods. In: Proceedings of the 23rd International Conference on Machine Learning (2006)
25. Wang, X., Mueen, A., Ding, H., Trajcevski, G., Scheuermann, P., Keogh, E.: Experimental comparison of representation methods and distance measures for time series data. Data Min. Knowl. Disc. **26**(2), 275–309 (2012)
26. Wei, L., Keogh, F.J.: Semi-supervised time series classification. In: Proceedings of KDD, pp. 748–753 (2006)
27. Weng, X., Shen, J.: Classification of multivariate time series using locality preserving projection. Knowl.-based Syst. **21**(7), 581–587 (2008)
28. Xing, Z., Pei, J., Keogh, E.: A brief survey on sequence classification. SIGKDD Explor. **12**(1), 40–48 (2010)
29. Yang, K., Shahabi, C.: A pca-based similarity measure for multivariate time series. In: Proceedings of the 2nd ACM International Workshop on Multimedia Databases, pp. 65–74 (2004)
30. Zhu, X.: Semi-supervised learning literature survey. Technical report TR 1530, University of Wisconsin Madison, Department of Computer Sciences (2008)
31. Zhu, X., Ghahramani, Z., Lafferty, J.: Semi-supervised learning using Gaussian fields and harmonic functions. In: Proceedings of the 20th International Conference on Machine Learning (2003)

Evaluating a Simple String Representation for Intra-day Foreign Exchange Prediction

Simon Cousins[✉] and Blaž Žličar[✉]

Department of Computer Science,
University College London, London WC1E 6BT, UK
{s.cousins,b.zlicar}@cs.ucl.ac.uk
http://www.ucl.ac.uk/

Abstract. This paper presents a simple string representation for hourly foreign exchange data and evaluates the performance of a trading strategy derived from it. We make use of a natural discretisation of the time-series based on arbitrary partitioning of the real valued hourly returns to create an alphabet and combine these individual characters to construct a string. The trading decision for each string is learnt in an incremental manner and is thus subject to temporal fluctuations. This naive representation and strategy is compared to the support vector machine, a popular machine learning algorithm for financial time series prediction, that is able to make use of the continuous form of past prices and complex kernel representations. Our extensive experiments show that the simple string representation is capable of outperforming these more exotic approaches, whilst supporting the idea that when it comes to working in high noise environments often the simplest approach is the most effective.

Keywords: Financial time series prediction · Discretisation · Parzen window estimators · Text kernels · Support vector machines

1 Introduction

The returns of financial time series (FTS) are renowned for being extremely noisy making it difficult to make predictions based upon these observations. This has led many previous authors to seek out novel and exotic representations of the time series that remove this noise and allow for better predictions to be made. In this paper we present an alternative view of the problem and introduce a simple discretisation operator on the observed returns. This discretisation, or binning, operator maps individual hourly returns from their original continuous form to a letter from an arbitrary, predefined alphabet. The letters corresponding to past returns are combined together to form a string representative of past price movements and we use this string to guide our trading decisions.

In this paper we examine 8 years of hourly data taken from four of the most actively traded currency pairs; AUD/USD, CHF/USD, EUR/USD and GBP/USD. The performance of our simple strategy is compared against support vector machines (SVMs), a popular machine learning algorithm for FTS

© Springer International Publishing Switzerland 2016
M. Ceci et al. (Eds.): NFMCP 2015, LNAI 9607, pp. 224–238, 2016.
DOI: 10.1007/978-3-319-39315-5_15

prediction. Extensive experiments are conducted using the SVM framework and a wide range of parameters are evaluated. The strategy that we present uses an incremental learning algorithm that records the average return of each unique market state representation (string). The decision on whether to trade is controlled by using (a) an expected return threshold and (b) a confidence proxy, measured in terms of the number of times we have visited that state. The results indicate that our relatively simple approach is capable of outperforming its more complicated counterpart, supporting our claim that the simplest solution is likely to fare best when in noisy environments.

2   Previous Work

This paper addresses the problem of FTS prediction from two aspects: (a) in terms of the complexity of an algorithm or a model and (b) in terms of data representation techniques.

A large part of FTS research focuses on pattern discovery within the noisy data. With regards to model selection, the usual approach is to use a type of autoregressive model so as to predict the future price or return trajectories. While this approach has proven quite useful for the prediction of mean-reverting and persistent processes such as volatility, it is less successful in predicting the value or even directional movement of a price time-series. Machine learning has demonstrated some encouraging results for the prediction of FTS, with SVMs being one of the most popular approaches. One of the first applications of SVMs to FTS was presented in [1] and later extended in [2]. Both papers provide an empirical analysis of SVM-based FTS prediction and compare its performance against a number of other techniques including multi-layer back-propagation neural network and case-based reasoning. The experimental results in both papers suggest a superiority of SVM methods when compared to similar techniques, however they do outline the challenges of the SVM approach in terms of generalisation. In [3] the authors compare the performance of Least Squares SVMs (LS-SVMs)[1] to that of several autoregressive models as well as non-parametric models for both return and volatility prediction. They report a superior performance of the LS-SVM in terms of achieving higher directional prediction and better overall performance compared to that of other models used in their investigation. Further applications of SVMs for financial forecasting can be found in [4–8].

In terms of data representation, the prevalent approach is to work with continuous time series data, favouring returns instead of prices due to a number of statistical properties that the former possess. Relatively less attention has been given to the study of the discretisation of financial time series, although there have been attempts to reduce the noise of a time series by using various quantisation techniques. Another way is to employ rule-based prediction methods that allow the incorporation of prior knowledge into the decision-making

---

[1] An extension of the original SVM that penalises the slack variables according to their squared value.

process. According to [9] rule-based forecasting involves two sources of knowledge (a) forecasting expertise (e.g. quantitative extrapolations and modelling) and (b) domain knowledge (practical knowledge about causal relations within particular field). Perhaps the most popular example of the latter in finance are methods where rules are based on technical indicators[2]. These allow a researcher to include their expert knowledge into the forecasting process in the form of various thresholds and patterns applied to technical indicators that ultimately lead to discrete evaluation of the market at a specific point in time. Understandably, rule-based forecasting goes beyond rules based on technical indicators, e.g. assigning different considerations to level and trend of a time series, combining predictions of a number of models, separating models according to their forecast horizons etc. [9].

In this sense, machine learning is less restrictive compared to classical time series analysis since it does not necessarily demand that the process satisfies a specific set of assumptions. While classical econometrics normally steers clear of including technical indicators and other heuristics, machine learning techniques allow for an easy inclusion of such indicators without violating any statistical assumptions. In fact it is quite popular to use these technical indicators as building blocks of the feature space when predicting FTS with machine learning algorithms. Moreover, it extends the possibilities of data representation through the use of kernel functions. The majority of implementations use the Gaussian kernel function due to its rich representation ability. Also frequently used are Polynomial and Laplacian kernels. To the best of our knowledge, text-based kernels have not yet been applied for pattern recognition in FTS, except in the context of news analytics. For more details on kernel methods see for example [10].

3   Simple Strategy

In this section we discuss a simple representation for FTS and explain how to use it to construct a trading strategy. At each hourly interval we have prices for the open, high, low and close over that interval, given by O_t, H_t, L_t and C_t respectively, with each price being a positive real number. Our goal is to predict at the beginning of the price interval whether the price will increase or decrease over the interval, therefore our target is $y_t = \text{sign}\,(C_t - O_t) \in \{-1, +1\}$. To do this we look at the relative price movements, or returns, $r_t = (C_t - O_t)/O_t$ that have occurred leading up to time t. Rather than using the continuous values of the returns we form a partitioning of the real line $\mathbb{R}$ and label each sub-interval with a unique identifier i.e. a letter σ from alphabet Σ.

The approach we take is to use sub-intervals $(b_k, b_{k+1}]$, where $b_k < b_{k+1}$, that have roughly an equal number of members, i.e. in the training sample there a roughly the same number of returns r_t corresponding to each sub-interval. This can be achieved by sorting the returns r_t and taking equally separated percentile

---

[2] Technical indicators are best described as rule-based evaluations of the underlying time series where their mathematical formulae is not based on statistical theory but on an expert's domain knowledge instead.

points as the limits of the sub-intervals. An important point to consider is the sub-interval that contains both positive and negative values. Intuitively it makes sense to split this sub-interval into two separate sub-intervals either side of zero.

The mapping $A : \mathbb{R} \to \Sigma$ between returns and the alphabet is given by

$$
A(r) = \begin{cases}
\sigma_1, & b_0 < r \leq b_1 \\
\sigma_2, & b_1 < r \leq b_2 \\
\quad\vdots & \\
\sigma_{|\Sigma|}, & b_{|\Sigma|-1} < r \leq b_{|\Sigma|},
\end{cases} \tag{1}
$$

where $b_0 = -\infty$ and $b_{|\Sigma|} = +\infty$. Using this notion each observation r_t belongs to a given sub-interval, which we identify with a letter σ from our alphabet Σ. The representation can be extended to include the past observations by concatenating the letters corresponding to past returns (see Fig. 1).

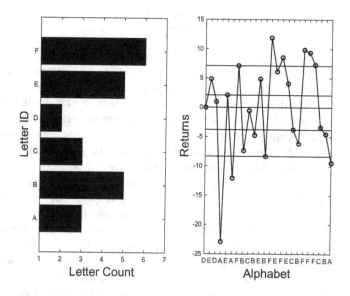

Fig. 1. Alphabet representation: left side displays letters (i.e. partitions) together with the number of examples (namely, the returns of EUR/USD exchange rate) that fall into individual partitions. Right side displays the time series of these same hourly returns over the period of 24 h with each of the hourly returns being assigned a letter depending on which partition they fall into.

More formally, an alphabet Σ of $|\Sigma|$ letters is constructed by an arbitrary partitioning of the real line $\mathbb{R}$. Each return $r_t \in \mathbb{R}$ is mapped to a letter $\sigma_t \in \Sigma$. A string is constructed according to $s_t = \sigma_{t-1} \ldots \sigma_{t-K}$, where K defines how many past returns we look at. Our goal is to come up with a prediction rule $g : \Sigma^K \to \{-1, 1\}$ indicating whether we believe the price will increase or decrease over the next interval. The feature space ϕ is constructed by mapping each unique string $s \in \Sigma^K$ to binary vector with one non-zero entry corresponding to that particular string such that $\langle \phi(s), \phi(s') \rangle = 1$ if and only if $s = s'$, and zero otherwise. This is equivalent to a bag-of-words kernel where we only have a single word in each document.

We learn our predictor in an incremental manner by maintaining an individual weight w_s and count c_s for each string. This weight w_s is given by the sum of observed outcomes for that string i.e. the weight corresponding to string s at time T is given by $w_s = \sum_{t=1}^{T} y_t I[s_t = s]$, where $I[a]$ is the indicator function returning 1 if the predicate a is true and 0 otherwise. The count c_s simply measures the number of times we have seen string s, i.e. $c_s = \sum_{t=1}^{T} I[s_t = s]$.

We can think of our predictor as an incremental version of the Parzen window classifier, which is traditionally used in conjunction with feature map ϕ and kernel function $k(x, x') = \langle \phi(x), \phi(x') \rangle$. The Parzen window prediction function g_{pw} is given by

$$g_{\mathrm{pw}}(x) = \mathrm{sign}\left(\sum_{t=1}^{T} y_t k(x_t, x) \right), \tag{2}$$

which takes into consideration each training example. This is often referred to as the Watson-Nadaraya estimator, which is an estimate of the conditional probability of a class. Note that we no longer have to maintain previous examples as they can be captured by the primal representation of the predictor and we only have to maintain $|\Sigma|^K$ individual weights. In our experiments this remains a feasible primal representation as the maximum alphabet length and string length are both set to 5 meaning that $|\Sigma|^K \leq 3125$. In Algorithm 1 we present the simple string based trading algorithm and have introduced two additional variables, a threshold τ and minimum observation number ν. The threshold τ can be interpreted as the excess in probability of a given class occurring that is required to invoke a trading decision. The minimum observation number ν is used to ensure that we can have gathered enough information in order to make a decision. Together these variables control the level of confidence that we have in our trading decision. The dimension of our primal weight vector $\mathbf{w}$ depends on the size of the alphabet Σ and the number of past returns K that we examine i.e. $|\mathbf{w}| = |\Sigma|^K$. At each time step t we only update a single entry of $\mathbf{w}$, the one corresponding to the particular string observation s_t at that time i.e. w_{s_t}. We construct and update the counts $\mathbf{c}$ in a similar manner.

4   String Subsequences Strategy

In this section we examine an extension of the simple string strategy by measuring the impact that subsequences of strings, known as n-grams, have on the probability of class membership, i.e. given a string s do we expect a price to go up or down during the following hour of trading. Furthermore, we introduce a new type of n-grams, Time decay n-grams, that as the name suggests take into account the conditional nature of FTS data.

Algorithm 1. Algorithm to run simple strings trading strategy

---

$SimpleStrings(S, R, \Sigma, K, \tau, \nu)$

Require: S-string representation, R returns, Σ alphabet, K time steps considered, τ trade threshold, ν minimum observation number

```
1:  Initialise weights w_s = 0 and counts c_s = 0 for each s ∈ Σ^K, Profit R = 0
2:  for t = 1 : T do
3:      Observe s_t, c_{s_t} compute f(s_t) = ⟨w, φ(s_t)⟩ = w_{s_t} and δ = w_{s_t}/c_{s_t}
4:      if δ > τ ∩ c_{s_t} > ν then
5:          Take long position (i.e. buy), p = 1
6:      else if δ < −τ ∩ c_{s_t} > ν then
7:          Take short position (i.e. sell), p = −1
8:      else
9:          Do not trade, p = 0
10:     end if
11:     Observe return r_t, y_t = sign(r_t)
12:     Update profit R ← R + pr_t
13:     Update observation counts c_{s_t} ← c_{s_t} + 1
14:     Update weight vectors w_{s_t} ← y_t + w_{s_t}
15: end for
```

4.1 n-Grams

Here we present a quick overview of the n-grams approach, a method for text representation popular especially in the fields such as computational linguistics [11] and bioinformatics [12]. One of the ways a text document can be represented is in terms of substrings where each substring represents a feature of the underlying document. As its name suggests n-grams refers to n-number of adjacent characters in the alphabet with each n-gram type representing a type of substring (i.e. a feature). We can write such a mapping of a document d_l into a vector space characterised by n-grams as $\phi : d_l \rightarrow \phi(d_l) \in F \subseteq \mathbb{R}^{|\Sigma|^n}$.

As an example let us consider that the entire document consists only of the word "excellent", (d_l = excellent) and we are interested in a 3-grams feature representation. The word "excellent" contains 7 unique 3-grams,

$$\text{excellent} \rightarrow [\text{ exc xce cel ell lle len ent }], \qquad (3)$$

which will correspond to 7 non-zero entries in the feature mapping ϕ. This simple example shows that the dimensionality in real life problems can increase very quickly. One of the challenges associated with n-grams is the choice of the value of n. In practice this value is normally relatively low as the dimensionality problem basically does not allow for very high values of n. In terms of our strings subsequence strategy, note that the size of the feature space is now $|\Sigma|^k$, where $k \leq K$ is the size of the subsequences that we examine. However our feature space is no longer a binary vector with one non-zero entry, instead there is a non-zero entry for each subsequence that is present in the string. The approach used in Algorithm 1 can be simply adjusted to account for the use of n-grams. The observation $s = (s_1^p \ldots s_k^p)$ is now decomposed to $k = \max(K - p + 1, 0)$

n-grams of length p, which results in $f(s) = \frac{1}{k}\sum_{i=1}^{k} w_{s_i^p}$. The weight vector and count updates occur in a synonymous manner to before, where we take each subsequence s_i^p in turn, updating its count $c_{s_i^p}$ and weight $w_{s_i^p}$. To keep our expressions as clear as possible we now drop the superscripts on s_i^p, when the context is clear that we use a fixed length n-gram. The confidence of the trading decisions are controlled by the total count $c = \frac{1}{k}\sum_i c_{s_i}$ and the value $\delta = f(s)/c$. The value δ can once again be interpreted as an estimate that the difference between the probability that price will increase versus decrease over the next time period, given that we have observed string s and all past strings.

4.2   Time Decay n-Grams

The n-gram feature space that we have described thus far corresponds to the traditional one used in machine learning literature, however this has not been designed to take into consideration any temporal influences that may exist. For example, we would expect that more recent subsequences will have a stronger influence on the likely outcomes and should therefore have a greater weight placed upon them. This empirical phenomenon is in financial literature often described as the conditional nature of FTS and is present across various asset classes and financial markets. To factor this into our representation we can introduce a simple decay function that weights subsequences according to their position in the string,

$$f(s) = \sum_{i=1}^{k} q_i w_{s_i} \text{ where } \sum_{i=1}^{k} q_i = 1 \text{ and } q_1 \geq q_2 \geq \cdots \geq q_k, \qquad (4)$$

note that in traditional n-grams we effectively have $q_i = \frac{1}{k}$ for each k. We have to make a slight adjustment to the updates, which are now given by $c_{s_i} \leftarrow q_i + c_{s_i}$ and $w_{s_i} \leftarrow q_i y + w_{s_i}$, where $\mathbf{w} = (w_s)_{s \in \Sigma^k}$ maintains a weighted sum of the directional movements associated with each subsequence. We now show that the prediction rule g is equivalent to that of a Parzen window classifier constructed using this new time decay feature mapping.

Proposition 1. *Let $\phi(s)$ be the feature mapping associated with the time-decay n-gram kernel of length k given by*

$$k(s, s') = \sum_{u \in \Sigma^k} \sum_{i=1}^{k} \sum_{j=1}^{k} q_i q_j I[s_i = u] I[s_j = u], \qquad (5)$$

where $\sum_{i=1}^{k} q_i = 1$ and $q_i \geq q_j$ if $i \geq j$. At time T the value of each component of the primal weight vector $(w_{s_i})_{s_i \in \Sigma^k}$ is given by $w_{s_i} = \sum_{t=1}^{T} y_t \sum_{i=1}^{k} q_i I[s_{t,i} = s_i]$, where $s_{t,i}$ corresponds to the i-th n-gram at time t. At time T the prediction function $g(s) = \text{sign}(\langle \mathbf{w}, \phi(s) \rangle) = \text{sign}\left(\sum_{i=1}^{k} q_i w_{s_i}\right)$ is equivalent to the Parzen window classifier given by $g_{pa}(s) = \text{sign}\left(\sum_{t=1}^{T} y_t k(s, s_t)\right)$.

Proof. By expressing the sum over kernel functions in terms of the subsequences present in string s we have

$$\sum_{t=1}^{T} y_t k(s, s_t) = \sum_{t=1}^{T} y_t \sum_{i=1}^{k} q_i \sum_{j=1}^{k} q_j I[s_{t,j} = s_i] = \sum_{i=1}^{k} q_i \sum_{t=1}^{T} y_t \sum_{j=1}^{k} q_j I[s_{t,i} = s_i],$$

(6)

which is equivalent to $\sum_{i=1}^{k} q_i w_{s_i}$. Therefore we see that both the Parzen window classifier and our simple average strategy will return the same prediction.

5   Experiments

We evaluate the performance of our proposed approach by using hourly data taken from four of the most actively traded currency pairs; AUD/USD, CHF/USD, EUR/USD and GBP/USD. The extensive dataset consists of 50,000 observations for each currency, covering dates ranging from February 2005 to January 2013. Digressing briefly, the majority of previous experiments using machine learning for FTS prediction have focused on predicting stock returns and often report abnormal returns. We have chosen to focus on currencies due to their permanence and relatively stable prices, rather than the survivorship bias and tendency for an upward drift in prices that exists with stocks taken from indices such as S&P500 or the FTSE100. We investigate the performance from two perspectives: (a) the type of data used for decision making and (b) the complexity of the learning algorithm. The evaluation is based solely on the trading criteria, namely on the cumulative profit accumulated by the underlying strategy over the 8 year trading period.

5.1   Simple String Strategy: Word vs. Alphabet Length

We begin by analysing the information contained within the discretised representation of daily returns and examine the performance of the simple strings strategy. Initially we run a simple strategy for a range of alphabet lengths $|\Sigma| = [2 : 6]$ and word lengths $K = [1 : 5]$. We set the minimum number of previously seen observations fixed at $\nu = 50$. The experiments are then conducted for two threshold values, $\tau = 0.00$ and $\tau = 0.05$. Figures 2 and 3 display the cumulative return for $\tau = 0.00$ and $\tau = 0.05$, respectively. Heat maps reflect the cumulative absolute return achieved by the simple strings strategy for various values of alphabet and word lengths. We see that strategies with short word and alphabet lengths achieve the best performance. This resembles the empirical fact that there is little to no autocorrelation between returns at higher frequencies and that the short look-back periods offer more information relative to longer periods. The performance does not seem to be overly sensitive to the size of the threshold τ, although the returns do seem to be slightly higher when $\tau = 0.05$ is used to assist the decision-making process. Enforcing a minimal conditional probability threshold therefore seems to assist trading decisions and in a way guarantees that the algorithm rules out at least some amount of noise.

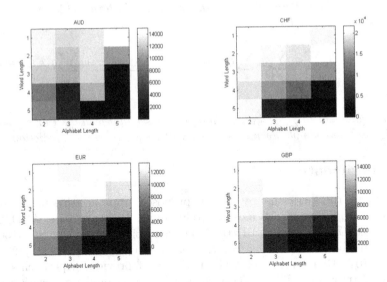

Fig. 2. Threshold $\tau = 0.00$: heat map shows a cumulative absolute return for the simple strategy for different alphabet and word lengths. The lighter the field the higher the absolute returns for a particular combination of alphabet and word lengths.

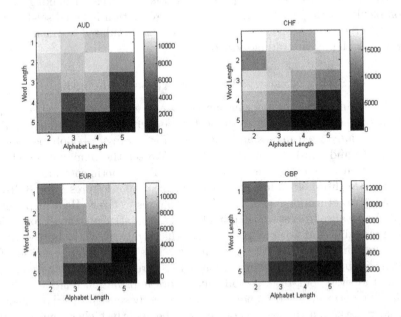

Fig. 3. Threshold $\tau = 0.05$: simple strategy applied with a non-zero trading threshold. Again, shorter alphabet and word lengths result in higher cumulative absolute returns.

5.2   Parzen Window Strategy: Regression vs. Classification

Here we investigate the information content of continuous hourly returns. While our goal is to predict the direction of each subsequent hourly return, i.e. $y_t \in \{-1, +1\}$, we can do so by means of either classification or regression. Consequently, we apply our incremental Parzen window approach as both a classification and a regression model and compare their performances for various parameterisations across all four currency pairs. We achieve the latter by simply changing the target from a binary value into a real-valued return and as such obtain an extension of the Parzen window regression (i.e. the Watson-Nadaraya regression). Figure 4 shows cumulative returns achieved for different combinations of word and alphabet lengths for both Parzen window classification (triangles) and regression (circles). These results suggest that classification method performs slightly better as it on average achieves higher cumulative returns. A possible explanation for this observation could be that the regression approach is much more sensitive to outliers and anomalous (infrequent and extremely high) returns, which affects the model long after the spike itself. The classification approach is less sensitive to such large price movements and weights each directional movement by the same amount. Consequently, the generalisation

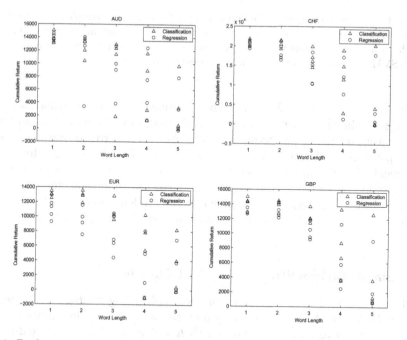

Fig. 4. Performance comparison between Parzen window classifier and Parzen window regression. Triangular (round) markers correspond to the returns generated by different combinations of world length, represented along x-axis and alphabet lengths, represented by different stars for the Parzen window classification (regression). Threshold $\tau = 0$ and minimum observation number $\nu = 50$.

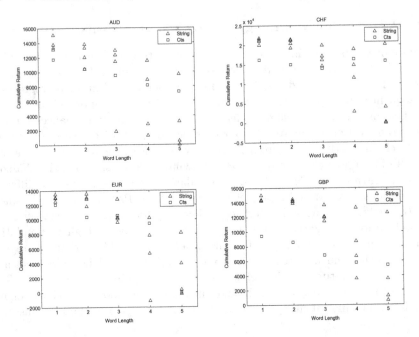

Fig. 5. Performance comparison between a linear SVM and the simple string representation. Triangular markers correspond to performance of different alphabet and word length string combinations for the simple string strategy and square markers are the linear SVM performance for continuous inputs of given length.

properties of the Parzen Window regression suffer while the classifier achieves better generalisation since the extreme price moves do not impact its solution, i.e. resulting in higher prediction accuracy and cumulative returns across currency pairs. Note that we can take into consideration the average directional movement of a currency by including a bias term i.e. $\phi(x) \leftarrow (1, \phi(x))$, however the experiments we conducted suggested it made no significant impact on performance. This can largely be explained by the tendency of currencies to revert to the original price. The results presented here are for experiments without the bias term.

5.3   Simple Strings Strategy vs. SVM Classification

Next, we compared the performance of our simple string strategy to the performance of a linear SVM classifier trained on the continuous representation of the respective string e.g. for a string of length $K = 5$ we used the last 5 real-valued returns to construct the input space for the SVM. Given the size of our dataset, we opted for a sliding window classifier approach, training for 1500 points, using a validation set of 500 points to choose the best regularisation parameter, and then testing the predictor on the subsequent 500 points. We can see in Fig. 5 that in general the simple string representation outperforms a linear SVM trained

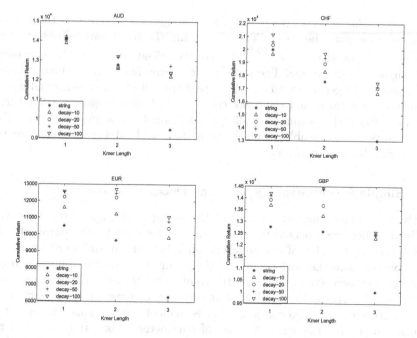

Fig. 6. Influence of time decay on performance. Minimum observation $\nu = 50$ and threshold $\tau = 0.01$. Reported are median returns for different model combinations with kmers of a given length (for different decay values). These results are then plotted against median performance of a simple string strategy using this length of string.

with continuous features. For most currencies and string parameterisations, the cumulative performance is higher than that of the continuous linear SVM. This results show that the simple discretisation and optimisation scheme can perform on par with more sophisticated learning algorithms. One explanation could be the fact the strength of the SVM comes from the generalisation guarantees offered by the margin that it obtains. However in particularly noisy situations, like FTS, the margin that the algorithm obtained is unable to provide reasonable generalisation guarantees. Furthermore the solution of the SVM is constructed based upon those examples that lie at the margin or on the wrong side of it, therefore we see that solutions in noisy situations are comprised of examples that lie on the periphery of the underlying distribution whereas the approach we have taken simply uses an efficient representation of the class conditional mean of these samples.

5.4 Time Decay n-Grams: Decay Analysis

We investigate the influence of the rate of decay on the performance of the time decay n-gram representation. We experimented using decay factors 10, 20, 50 and 100 for three different kmer lengths. To compute the weightings q_i used with the function $f(s) = \sum_i q_i w_i$ we begin by letting $\tilde{q}_1 = 1$ and recursively compute

$\tilde{q}_{i+1} = \tilde{q}_i c$. The decay factor d states that $\tilde{q}_1/\tilde{q}_k = d$, where k is the number of kmers present in the string. Therefore using the recursions we observe that $\tilde{q}_1 = d\tilde{q}_1 c^{k-1}$ and that $c = e^{-\frac{\log d}{k-1}}$. A final normalisation $q_i = \tilde{q}_i / \sum_j \tilde{q}_j$ ensures the weights q_i sum to one. The results of the experiment quite clearly show a preference for larger decay factors. This perhaps reflects the conditional nature of time series returns and the weak correlation of high-frequency FX returns. Moreover, Fig. 6 shows that by adding a conditional characteristic to the string representation we are able to outperform the simple string approach for the majority of decay parameters.

5.5   Simple String Strategy vs. Time Decay n-Grams

Lastly, to evaluate the performance of the string subsequences representation outlined earlier, we repeated the same experiments used for the simple strings, whilst varying the size of the n-grams, alphabet length and word length. In Fig. 7 we compare the performance of the string subsequence approach to our previous approach. We can see from the results that the n-gram implementation performs similarly to the original string approach, however it appears to be less sensitive to parameterisation. This is evident from a smaller amount of variance in the returns across a range of parameters. Note, that for all of the

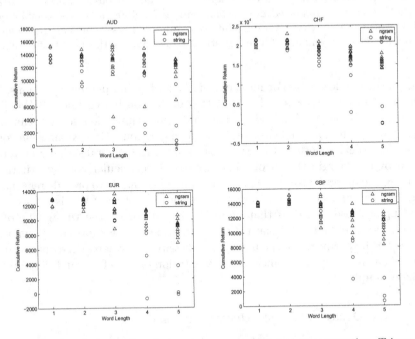

Fig. 7. Performance comparison between n-gram and string representation. Triangular markers correspond to performance of different n-gram combinations of n-gram length and alphabet length. Circles are the string performance for different alphabet and word lengths.

experiments presented, we purposely used a simple decay factor of 10 (meaning that $q_1/q_K = 10$), which is the value that in previous section performed the worst and we still record a performance as good or better than the one by simple string strategy. This shows great promise for the time-decay subsequence approach and something that could be studied more extensively in the future.

6 Conclusions

This paper presents a novel approach to FTS forecasting based on a simple string representation. We examine the design of a trading strategy from two perspectives (a) the complexity of the underlying algorithm and (b) the representation of the underlying time series used in the decision making process. We compare a simple approach based on discrete data to the popular linear SVM with continuous inputs, and introduce a new type of text kernel (Time decay n-grams), that captures temporal influences of string subsequences. Furthermore we show that this kernel can be evaluated efficiently using a simple weighted averaging process that is equivalent to the Parzen window classifier using that kernel. The results of these experiments suggest that a simple string representation coupled with an averaging process is capable of outperforming more exotic approaches, whilst supporting the idea that when it comes to working in high noise environments often the simplest approach is the most effective.

References

1. Tay, F.E.H., Cao, L.: Application of support vector machines in financial time series forecasting. Omega **29**, 309–317 (2001)
2. Kim, K.J.: Financial time series forecasting using support vector machines. Neurocomputing **55**, 307–319 (2003)
3. Van Gestel, T., Suykens, J.A.K., Baestaens, D.E., Lambrechts, A., Lanckriet, G., Vandaele, B., De Moor, B., Vandewalle, J.: Financial time series prediction using least squares support vector machines within the evidence framework. Neural Netw. **12**, 809–821 (2001)
4. Cao, L.J., Tay, F.E.H.: Support vector machine with adaptive parameters in financial time series forecasting. Neural Netw. **14**, 1506–1518 (2003)
5. Perez-Cruz, F., Afonso-Rodriguez, J.A., Giner, J.: Estimating GARCH models using support vector machines. Quant. Finan. **3**, 163–172 (2003)
6. Hossain, A., Nasser, M.: Reccurent support and relevance vector machines based model with applications to forecasting volatility of financial returns. J. Intell. Learn. Syst. Appl. **3**, 230–241 (2011)
7. Ou, P., Wang, H.: Financial volatility forecasting by least square support vector machine based on GARCH, EGARCH and GJR models: evidence from ASEAN stock markets. Int. J. Econ. Finan. **2**, 51–64 (2010)
8. Khan, A.I.: Financial volatility forecasting by nonlinear support vector machine heterogeneous autoregressive model: evidence from NIKKEI225 stock index. Int. J. Econ. Finan. **3**, 138–150 (2011)
9. Armstrong, J.S., Adya, M., Collopy, F.: Rule-Based Forecasting: Using Judgment in Time-Series Extrapolation. Kluwer Academic Publishers, Norwell (2001)

10. Shawe-Taylor, J., Cristianini, N.: Kernel Methods for Pattern Analysis Infrastructure. Cambridge University Press, New York (2004)
11. Lodhi, H., Saunders, C., Shawe-Taylor, J., Cristianini, N., Watkins, C.: Text classification using string kernels. J. Mach. Learn. Res. **2**, 419–444 (2002)
12. Liu, B., Wang, X., Lin, L., Dong, Q., Wang, X.: A discriminative method for protein remote homology detection and fold recognition combining top-n-grams and latent semantic analysis. BMC Bioinf. **9**, 510 (2008)

Author Index

Printed in the United States
By Bookmasters